Björn Schmolck

Omitted Variable Tests and Dynamic Specification

An Application to Demand Homogeneity

Springer

Author

Dr. Björn Schmolck
Swiss Reinsurance Company
Economic Research & Consulting
Mythenquai 50/60
CH-8022 Zurich, Switzerland

E-mail: Bjoern_Schmolck@swissre.com

Library of Congress Cataloging-in-Publication Data

Schmolck, Björn, 1968-
    Omitted variable tests and dynamic specification : an application to demand
homogeneity / Björn Schmolck.
        p. cm. -- (Lecture notes in economics and mathematical systems, ISSN 0075-8442 ; 488)
    Thesis (Ph.D.)--University of Fribourg, Switzerland, 2000?
    Includes bibliographical references.
    ISBN 978-3-540-67358-3   ISBN 978-3-642-58324-7 (eBook)
    DOI 10.1007/978-3-642-58324-7
    1. Demand (Economic theory) 2. Time-series analysis. 3. Regression analysis. I. Title.
II. Series.

    HB842 .S36 2000
    519.5'36--dc21

                                                                    00-038753

ISSN 0075-8442
ISBN 978-3-540-67358-3

Typesetting: Camera ready by author
Printed on acid-free paper   SPIN: 10766690   42/3143/du-543210

Lecture Notes in Economics and Mathematical Systems

488

Springer-Verlag Berlin Heidelberg GmbH

Preface

This book deals with the omitted variable tests for a multivariate time-series regression model. What are the consequences of testing for the omission of a variable when the model is dynamically misspecified? What is the small sample bias of the omitted variable test when the model dynamics is correctly specificfied? The answers to these questions are proposed in this book. As an empirical illustration, the analysis is applied to the homogeneity test of a demand system.

I particularly thank Professor Dr. Philippe J. Deschamps who draw my attention on this subject and who made very helpful comments and suggestions. Additionally, I would like to thank Professor Dr. Reiner Wolff for his comments especially on the chapter dealing with consumer theory. Special thanks go to María José Redondo, who read this book several times and for the inspiring discussions with her. I would also like to thank Dr. Ali Vakili (always ready to answer any questions in mathematics), Prof. Dr. Hans Wolfgang Brachinger, Curzio De Gottardi, Peter Mantsch, Dr. Paul-André Monney, Dr. Uwe Steinhauser, Leon Stroeks and Dr. Peter Windlin. Frances Angell improved the English of this work.

The research for this book had been financially supported by the Université de Fribourg (Switzerland).

Finally, I appreciated the support from Springer-Verlag and I thank Dr. Werner A. Müller and Ruth Milewski for their friendly collaboration. An anonymous referee from Springer-Verlag made helpful comments.

Abstract

The purpose of this work is to study the omitted variable test for a multivariate regression model. The empirical motivation for this is the homogeneity test for a demand system; the hypothesis of homogeneity can be formulated as the hypothesis of an omitted variable.

The exact distribution of the test statistic for homogeneity is only known for static demand systems. The static models are, however, very restrictive for time-series models. It is, therefore, interesting to study the consequences of dynamic misspecification for the omitted variable test. In most of our examples the homogeneity test is biased substantially towards rejection. This result illustrates the importance of specifying the dynamics of the demand system correctly.

In order to take the dynamics of the demand system into account, we analyse two classes of dynamically correctly specified tests for homogeneity: robust Wald tests and versions of the likelihood ratio test. Since the null distribution of the related test statistics is only known asymptotically, the small sample performance of these tests is studied by Monte Carlo experimentation. In our examples, the robust Wald test and the usual likelihood-ratio test perform badly and the null hypothesis of homogeneity is rejected too often in most of the cases. As a remedy we propose a bootstrap version of the likelihood ratio test which has an excellent performance.

Table of Contents

1. Introduction

This book deals with the omitted variable test for the multivariate time-series regression model. The empirical motivation for this is the homogeneity test for a demand system; the hypothesis of homogeneity can be formulated as the hypothesis of an omitted variable. We are interested in studying the omitted variable test under dynamic misspecification and under correct specification.

The exact distribution of the test statistic for homogeneity is only known for static demand systems. This is Laitinen's statistic which can be interpreted as the F-version of the Wald statistic. Since the static models are very restrictive for time-series models, it is interesting to study the consequences of dynamic misspecification for the omitted variable test.

The analysis is applied to a restricted form of dynamic models which facilitates the interpretation of the results. We are interested in the omitted variable test for the multivariate linear regression model of the form $y_t = Bx_t + u_t$, where y_t is a $n \times 1$ vector of endogenous variables, B is a $n \times k$ coefficient matrix, x_t is a $k \times 1$ vector of exogenous regressors and u_t is a $n \times 1$ vector of serially correlated errors. The null hypothesis is that one column of the coefficient matrix B equals zero. This is equivalent to testing that one variable can be omitted. Many empirical applications can be analysed in this framework. Here we will study the homogeneity test for a demand system. The demand system under consideration is the Rotterdam model. More generally, an omitted variable test for the reduced form of an equation system fits into this framework.

For univariate regression models, the distributions of some test statistics have been analysed when the errors are dynamically misspecified. For the multivariate regression model, however, the omitted variable test has not yet been analysed under dynamic specification to our knowledge. The test under consideration is the F-version of the Wald test for a multivariate regression model. This test is exact if the regressors are exogenous and the errors are normal, homoskedastic and independent in time. We study the type I error of this Wald test when the error process is dynamically misspecified, for example if the errors follow a vector autoregressive or moving average process. It will be shown that the Wald statistic is distributed asymptotically as a quadratic form in normal variables under quite general assumptions. The asymptotic distribution depends on the data-generating process and its functional form is based on Imhof's formula.

In order to illustrate the bias of the omitted variable test under dynamic misspecification, some examples will be presented. Here, the interest is to study the bias of Laitinen's test under dynamic misspecification. For some specific data-generating processes the true type I error is estimated by simulation. The true data-generating processes are based on the homogeneity constrained estimation of the Rotterdam model with annual and seasonal data from the UK. The errors are assumed to follow a vector autoregressive or moving average process. In our examples, the homogeneity test is biased substantially towards rejection. The estimated rejection frequencies vary between 4% and 46% for a given nominal type I error of 5%. This illustrates the importance of using a dynamically correctly specified test for homogeneity. It is, furthermore, one plausible reason why homogeneity has been rejected in many empirical applications.

We analyse the small sample performance of two classes of correctly specified tests for homogeneity: robust Wald tests and versions of the likelihood ratio test. The small sample performance of these tests will be analysed by Monte Carlo experimentation.

The class of *robust Wald tests* is based on the "heteroskedasticity and autocorrelation consistent" (HAC) variance-covariance matrix estimators which have become quite popular in the recent years. These Wald tests are robust in the sense that the Wald statistics have an asymptotic chi-square distribution under fairly general conditions. The small sample performance of these tests will be investigated by simulation. In the literature, the small sample properties of these robust tests have only been studied for univariate and not for multivariate regression models to the author's knowledge. Monte Carlo experimentation suggests that the bias of these tests is extremely large and increases with the number of equations of the model. The small sample properties of these tests, therefore, differ for univariate and multivariate regression models. The conclusion from the Monte Carlo experiments is that the robust Wald tests should not be used when testing demand homogeneity in a time-series models if working with annual or seasonal data.

The other class of tests for homogeneity is related to the *likelihood ratio test*. In order to take the small sample problem into account, two possible remedies are proposed: a modified likelihood ratio statistic with a small sample correction factor and a Monte Carlo test. In the simulation study we have found that the likelihood ratio test does not perform well. The small sample corrected version of this test works better but not satisfactorily. The bias increases dramatically with the number of equations. The performance of the Monte Carlo test, however, is excellent and, therefore, recommended.

Organisation of the book:

In chapter 2, a simple model is presented in order to study the omitted variable test under dynamic misspecification. The model is a linear regression model with one autoregressive regressor and an autoregressive error term. The asymptotic and small sample properties of the misspecified t-test are analysed. This example illustrates some applications of asymptotic theory and Monte Carlo experimentation in a simple framework. It can be seen as an introduction to what follows in the rest of the book.

In the following chapters, the omitted variable test for a multivariate regression model is studied under incorrect and correct dynamic specification. The analysis is applied to the Rotterdam model.

In chapter 3, some basic concepts of consumer theory are presented in order to describe the theoretical background of the Rotterdam model.

Chapter 4 deals with the robust estimation of the coefficients of the Rotterdam model and its variance-covariance matrix. For estimating the coefficients of the multivariate regression model, the quasi-maximum likelihood estimator is derived. This estimator is consistent for a broad class of error processes. Furthermore, some robust estimators of the variance-covariance matrix of the quasi-maximum likelihood estimator are presented; these are the so-called "heteroskedasticity and autocorrelation consistent" (HAC) variance-covariance matrix estimators. We consider the spectral kernel estimator proposed in Andrews (1991), the prewhitened spectral kernel estimator (Andrews and Monahan, 1992) and the so-called "VARHAC" estimator proposed in Den Haan and Levin (1997).

In chapter 5, various tests for homogeneity will be discussed. We consider Anderson's $\mathcal{U}$ test if the errors are normal homoskedastic and time-independent. It will be shown that Anderson's $\mathcal{U}$ statistic is functionally equivalent to Laitinen's statistic. Although this equivalence is mentioned in the literature, its demonstration has never been given explicitly to the authors's knowledge. Then the likelihood ratio test for homogeneity for a demand system with vector autoregressive errors is defined. In order to take the small sample problem into account, two related likelihood ratio tests are proposed. These are a small sample corrected likelihood ratio test based on Anderson's $\mathcal{U}$ test and a Monte Carlo test. In this chapter, we also study the consequences of testing for homogeneity when the error process is wrongly specified. This is the main analytical contribution of this work. Finally, various robust Wald tests for homogeneity are defined. These Wald tests are based on the quasi-maximum likelihood estimation of the coefficients of the model and the heteroskedasticity and autocorrelation consistent variance-covariance matrix estimators. A small sample corrected version of these tests is also proposed.

In chapter 6, the small sample properties of the various homogeneity tests discussed in chapter 5 are studied by simulation. The simulation analysis can be seen as an investigation of the omitted variable test for a linear multivariate regression system with dynamic errors and a specific data-generating process. In order to define some "realistic" data-generating processes, the population parameters of these processes are derived from the Rotterdam model estimated with annual and seasonal data from the UK. The experimental design is motivated and described in detail. Finally, the simulation results are discussed.

Chapter 7 concludes.

2. The t-statistic under dynamic misspecification

In this chapter, we analyse the t-statistic for a linear model with one stochastic regressor and with autocorrelated errors under dynamic misspecification. The motivation for this is as follows. The t-test for a univariate regression model with one regressor is a special case of an omitted variable test for a multivariate regression model with multiple regressors. The homogeneity test for a consumer demand system can be formulated as an omitted variable test. This will be studied in detail in the chapters 3-6. The example studied in this chapter illustrates some applications of asymptotic theory and Monte Carlo experimentation in a simple framework. It can be seen as an introduction to what follows in the other chapters.

This chapter is organised as follows: in section 2.1, the model is defined. In section 2.2, the properties of the ordinary least-squares (OLS) estimators and in section 2.3, the asymptotic distribution of the t-statistic are derived. In section 2.4, the principle of Monte Carlo experimentation is described and applied to the t-test in section 2.5.

2.1 The model

The following assumptions are made:

Assumption 2.1

a) The model is given by $y_t = \beta x_t + u_t$ for $t = 1, 2, \cdots, T$. In matrix notation: $y = \beta x + u$, where y, x, u are column vectors and β is a scalar.
b) x_t is a stochastic scalar variable which is generated by an autoregressive process of order 1 (AR(1)): $x_t = \phi x_{t-1} + \mu_t$ with $\mu_t \sim i.i.d.N(0; \sigma_\mu^2)$ and $|\phi| < 1$, where $i.i.d.$ means independent and identical distributed.
c) The disturbance term u_t is also generated by an AR(1) process: $u_t = \rho u_{t-1} + \epsilon_t$ with $\epsilon_t \sim i.i.d.N(0; \sigma_\epsilon^2)$ and $|\rho| < 1$.
d) x and u are independent.

Assumptions 2.1 b) and c) imply that the series x_t and u_t are covariance-stationary. This means that the mean and covariances of each series do not depend on date t (Hamilton 1994, p.45ff):

$E(x_t) = 0$ and $E(u_t) = 0$.

The variances and covariances are given by

$$\sigma_x^2 := E(x_t^2) = \frac{\sigma_\mu^2}{1 - \phi^2} \quad \text{and} \quad Cov(x_{t-i}, x_{t-j}) = \phi^{|i-j|}\sigma_x^2 \quad ,$$

$$\sigma_u^2 := E(u_t^2) = \frac{\sigma_\epsilon^2}{1 - \rho^2} \quad \text{and} \quad Cov(u_{t-i}, u_{t-j}) = \rho^{|i-j|}\sigma_u^2 \quad .$$

The variance-covariance matrix of x is given by

$$\Sigma_x := E(xx') = \sigma_x^2 \Omega_x$$

with

$$\Omega_x = \begin{bmatrix} 1 & \phi & \phi^2 & \cdots & \phi^{T-1} \\ \phi & 1 & \phi & \cdots & \phi^{T-2} \\ \phi^2 & & \ddots & & \vdots \\ \vdots & & & \ddots & \vdots \\ \phi^{T-1} & \cdots & \cdots & \cdots & 1 \end{bmatrix} \quad .$$

We denote the element of Ω_x in the ith row and the jth column by $[\Omega_x]_{ij}$. The variance-covariance matrix of u is given by

$$\Sigma_u := E[uu'] = \sigma_u^2 \Omega_u \quad ,$$

with $[\Omega_u]_{ij} = \rho^{|i-j|}$.

2.2 Properties of the estimators

In this section, the properties of the OLS estimator for β and the usual residual variance estimator are derived. The results of this section are summarised by the following proposition:

Proposition 2.1

Given assumption 2.1, the following holds:

a) The estimator $b := (x'x)^{-1}x'y$ is unbiased.
b) The estimator $b := (x'x)^{-1}x'y$ is consistent.
c) The estimator $s_u^2 := \hat{u}'\hat{u}/(T-1)$ with $\hat{u} := y - bx$ is consistent.

Proof of Proposition 2.1 a): Unbiasedness of b

It will be proven that

$$E(b) = \beta \quad .$$

The estimator b can be written as

$$b := (x'x)^{-1}x'y = (x'x)^{-1}x'(\beta x + u) = \beta + (x'x)^{-1}x'u \quad .$$

As the regressor x is stochastic, the expectation of b conditional on x is calculated first:

$$E(b|x) = \beta + (x'x)^{-1}x'E(u|x) = \beta \quad .$$

Therefore, b is an unbiased estimator of β conditional on x.
Using the law of iterated expectations and assumption 2.1 d), one can see that b is unbiased although x is stochastic:

$$E(b) = E_x[E(b|x)] = \beta + E_x[(x'x)^{-1}x'\underbrace{E(u|x)}_{=0}] = \beta \quad .$$

$E_x(.)$ signifies that the expectation is taken over the distribution of x.

Proof of Proposition 2.1 b): Consistency of b

It can be shown that b is a consistent estimator [1] of β. Multiplying and dividing the estimator b by T yields:

$$b = \beta + \left(\frac{x'x}{T}\right)^{-1}\frac{x'u}{T} \quad .$$

The scalar term $x'x/T$ converges in probability to σ_x^{-2} and $x'u/T$ to zero. This is shown first for $x'x/T$ by verifying the following two conditions [2] :

$$\lim_{T\to\infty} E\left(\frac{x'x}{T}\right) = \sigma_x^2 \text{ and } \lim_{T\to\infty} V\left(\frac{x'x}{T}\right) = 0.$$

The expectation and the variance of a quadratic form with $x \sim N(0, \Sigma_x)$ can be calculated with a simplified version of a theorem [3] given in Magnus and Neudecker (1988, p.251, theorem 12) in the following way:

---

[1] In general an estimator $\hat{\theta}$ of θ is said to be consistent if $\hat{\theta}$ converges in probability to the constant θ: $\hat{\theta} \xrightarrow{P} \theta$. This means that

$$\lim_{T\to\infty} P(|\hat{\theta}_T - \theta|) > \epsilon) = 0$$

for any $\epsilon > 0$ (Amemiya 1985, p.85 and p.95).

[2] Often, it is easier to demonstrate convergence in probability of the estimator $\hat{\theta}$ by verifying the following two conditions:
$\lim_{T\to\infty} E(\hat{\theta}) = \theta$ and $\lim_{T\to\infty} Var(\hat{\theta}) = 0$. These conditions are sufficient but not necessary for convergence in probability. The two conditions are an implication of the convergence in mean square. The condition is only sufficient as convergence in mean square implies convergence in probability but not vice versa (compare Amemiya 1985, p.85ff).

[3] The theorem is (Magnus and Neudecker 1988, p.251, theorem 12): If x is $N(\mu, \Omega)$ and A is a symmetric $n \times n$ matrix, then

$$E(x'Ax) = \text{tr } A\Omega + \mu'A\mu$$

and

$$Var(x'Ax) = 2\text{tr } (A\Omega)^2 + 4\mu'A\Omega A\mu. \quad .$$

In our example, $A = I$, $\mu = 0$ and $\Omega = \sigma_x^2\Omega_x$.

$E(x'x) = \operatorname{tr} \Sigma_x$ and $Var(x'x) = 2\operatorname{tr} \Sigma_x^2$.
Using this theorem and $\operatorname{tr} \Omega_x = T$ yields

$$E\left(\frac{x'x}{T}\right) = \frac{\sigma_x^2}{T} \operatorname{tr} \Omega_x = \sigma_x^2 \quad.$$

The variance of $x'x$ is:

$$Var(x'x) = 2\sigma_x^4 \operatorname{tr} \Omega_x^2 = 2\sigma_x^4 \sum_{i=1}^{T}\sum_{j=1}^{T}[\Omega_x]_{ij}^2$$

$$= 2\sigma_x^4 \left(T + 2(T-1)\phi^2 + 2(T-2)\phi^4 + \cdots + 2\phi^{2(T-1)}\right) \quad.$$

The trace of Ω_x^2 has been calculated by using the rule $\operatorname{tr} A'B = (\operatorname{vec} A)' \operatorname{vec} B$, where A and B are matrices of the same order. Using the fact that Ω_x is symmetric and all elements on a specific diagonal are equal (this is a symmetric Toeplitz matrix), the result follows.

Noting that $|\phi| < 1$ by assumption, it can be seen that the last expression is at most of order T denoted by $O(T)$ which implies that $\lim_{T\to\infty} \frac{1}{T^2} Var(x'x) = 0$. Therefore,

$$\frac{x'x}{T} \xrightarrow{P} \sigma_x^2 \quad.$$

When calculating the probability limit of $\left(\frac{x'x}{T}\right)^{-1}$, one can apply Slutsky's theorem (Amemiya 1985, p.89) [4] which yields:

$$\left(\frac{x'x}{T}\right)^{-1} \xrightarrow{P} (\sigma_x^2)^{-1} \quad.$$

Next, it is shown that $\frac{x'u}{T} \xrightarrow{P} 0$. The conditional expectation is:

$$E\left(\frac{x'u}{T}\Big|x\right) = 0 \quad.$$

By using the law of iterated expectation one can see that the unconditional expectation is also zero:

---

[4] Let x_T be a sequence of (vector) random variables and $g(.)$ be a real-valued function continuous at a constant point α. If $x_T \xrightarrow{P} \alpha$, then $g(x_T) \xrightarrow{P} g(\alpha)$. A special case of this theorem is the following: If $x_T \xrightarrow{P} c$ and $y_T \xrightarrow{P} d$, then

$$\text{i)} \quad x_T + y_T \xrightarrow{P} c + d \quad \text{and} \quad \text{ii)} \quad x_T y_T \xrightarrow{P} cd \quad.$$

$$E\left(\frac{x'u}{T}\right) = E_x\left[E\left(\frac{x'u|x}{T}\right)\right] = 0 \quad .$$

The conditional variance of $x'u$ is:

$$Var\left(x'u|x\right) = x'E\left(uu'|x\right)x = \sigma_u^2 x'\Omega_u x \quad .$$

The unconditional variance can be found by decomposing the variance:

$$Var(x'u) = E_x[Var(x'u|x)] + V_x[E(x'u|x)] \quad ,$$

where $V_x(.)$ signifies that the variance is taken over the distribution of x. Using the theorem given on page 7 (footnote 3) yields:

$$
\begin{aligned}
Var(x'u) &= \sigma_u^2 E_x[x'\Omega_u x] + 0 = \sigma_u^2 tr(\sigma_x^2 \Omega_u \Omega_x) \\
&= \sigma_u^2 \sigma_x^2 \left(T + 2(T-1)\rho\phi + 2(T-2)(\rho\phi)^2 + \cdots + 2(\rho\phi)^{2(T-1)}\right).
\end{aligned}
$$

The expression for $Var(x'u)$ is O(T) and therefore

$$\lim_{T\to\infty} \frac{1}{T^2} Var(x'u) = 0 \quad .$$

When calculating the probability limit of b one can use Slutsky's theorem. Therefore,

$$b \xrightarrow{p} \beta + (\sigma_x^2)^{-1} \cdot 0 = \beta \quad .$$

Proof of Proposition 2.1 c): Consistency of s_u^2

The variance of the residuals is estimated in the "naive" way which means that the autocorrelation of the errors is ignored. It will be shown that this estimator is consistent.
The estimator of the residual variance is defined as

$$s_u^2 := \frac{\hat{u}'\hat{u}}{T-1} \quad .$$

The residual vector $\hat{u}$ can be rewritten by using its definition [5]:

$$\hat{u} = [I - x(x'x)^{-1}x']u = M_x u \quad .$$

Using the definition of M_x and the fact that M_x is symmetric and idempotent[6], s_u^2 can be approximated by

$$s_u^2 \approx \frac{u'M_x u}{T} = \frac{u'u}{T} - \frac{u'x}{T}\left(\frac{x'x}{T}\right)^{-1}\frac{x'u}{T} \quad ,$$

---

[5] $\hat{u} = y - \hat{y} = y - xb = y - x(x'x)^{-1}x'y = \underbrace{[I - x(x'x)^{-1}x']}_{=M_x}u = M_x u$

[6] A square matrix A is said to be idempotent if $AA = A$.

where $T - 1$ was replaced by T since we are interested in the probability limit. The convergence in probability of $u'u/T$ can be demonstrated in the same way as for $x'x/T$. Therefore,

$$\frac{u'u}{T} \overset{P}{\to} \sigma_u^2 \quad . \tag{2.1}$$

As shown before, $x'u/T \overset{P}{\to} 0$ and $(x'x/T)^{-1} \overset{P}{\to} (\sigma_x^2)^{-1}$. The consistency of s_u^2 can be shown by using Slutsky's theorem, which establishes that s_u^2 is a consistent estimator of σ_u^2:

$$s_u^2 \overset{P}{\to} \sigma_u^2 \quad .$$

2.3 The distribution of the quasi t-statistic

Here a usual OLS t-test is applied to test if β is different from zero. This hypothesis can be formulated as

$$H_0: \quad \beta = 0 \quad \text{and} \quad H_1: \quad \beta \neq 0 \quad .$$

As the regressor x is stochastic and the errors are autocorrelated, the exact distribution of the test statistic does not follow a student distribution in general. Therefore, the asymptotic distribution of this test will be derived.

The quasi t-statistic is defined by

$$t_T := \frac{b}{\sqrt{s_u^2 (x'x)^{-1}}} \quad .$$

Proposition 2.2

Given assumption 2.1, the following holds:

a) The quasi t-statistic has a student distribution if ρ or ϕ equals zero.
a) The asymptotic distribution of the quasi t-statistic is given by

$$t_T \overset{d}{\to} N\left(0, \frac{1 + \rho\phi}{1 - \rho\phi}\right) \quad .$$

Proof of proposition 2.2a)

The errors are spherical if $\rho = 0$. For this case, it is known that t_T is exactly distributed as a student with $T - 1$ degrees of freedom if the regressor x is non-stochastic. Here the regressor x is stochastic and, therefore, t_T follows a student distribution conditional on x. The t-distribution is, however, completely determined by the degrees of freedom of the random variable t_T

(Hogg and Craig 1978, p.182). This means that it does not depend on the stochastic regressor x. Therefore, the unconditional distribution of t_T is also distributed as a student if $\rho = 0$.

This result also holds for $\phi = 0$. This can be seen by rewriting t_T in the following way:

$$t_T \overset{H_0}{=} \frac{(x'x)^{-1}x'u}{\sqrt{\frac{u'M_x u}{(T-1)}}(x'x)^{-1}} = \frac{\sqrt{(T-1)}(x'u)}{\sqrt{u'ux'x - (u'x)^2}} = \frac{(u'u)^{-1}u'x}{\sqrt{\frac{x'M_u x}{(T-1)}}(u'u)^{-1}} \quad,$$

(2.2)

with $M_u = \left[I - u(u'u)^{-1}u'\right]$. This reformulation only involved simple manipulations, keeping in mind that $x'x$, $u'u$ and $x'u$ are scalars. Therefore, we can conclude that the t-statistic follows exactly a student distribution if ρ or ϕ equal zero.

Proof of proposition 2.2b)

Here two alternative proofs are given.

First proof:

Rewriting t_T yields:

$$t_T = s_u^{-1}\left(\frac{x'x}{T}\right)^{-\frac{1}{2}}\frac{x'u}{\sqrt{T}} \quad.$$

The first two terms on the right side of the equation are converging in probability to $\sigma_u^{-1}\sigma_x^{-1}$. The second term converges in distribution [7]. For establishing the latter result, one can use a central limit theorem for a serially correlated sequence which can be presented in the following form (see Hamilton 1994, p.195 or Anderson 1971, p.429):

If

$$a_t = \sum_{j=0}^{\infty}\psi_j v_{t-j} \quad,$$

where v_t is sequence of i.i.d. random variables with $E(v_t) = 0, E(v_t^2) < \infty$ and $\sum_{j=0}^{\infty}|\psi_j| < \infty$, then

$$\frac{\sum_{t=1}^{T}a_t}{\sqrt{T}} \overset{d}{\to} N(0, \sum_{j=-\infty}^{\infty}\gamma_j)$$

(2.3)

with $\gamma_j := Cov(a_t, a_{t-j})$.

---

[7] A sequence X_T is said to converge to X in distribution if the distribution function F_T of X_T converges to the distribution function F of X at every continuity point of F (Amemiya 1985, p.85).

Let $a_t = x_t u_t$. One can show that a_t is a covariance-stationary sequence:
i)$E(a_t) = E(x_t u_t) = 0$ for all t by using the law of iterated expectations.
ii)$Cov(a_t, a_{t-j}) = E(x_t u_t x_{t-j} u_{t-j}) = E_x[x_t x_{t-j} E[u_t u_{t-j}|x_t x_{t-j}]] = \rho^j \phi^j \sigma_u^2 \sigma_x^2$
for all t and any j.
By Wold's decomposition theorem (Hamilton, p.109), one can write any zero-mean covariance stationary process as

$$a_t = \sum_{j=0}^{\infty} \psi_j v_{t-j} + \kappa_t \quad , \qquad (2.4)$$

where $\psi_0 = 1$ and $\sum_{j=0}^{\infty} \psi^2 < \infty$. v_t is white noise and defined as

$$v_t := a_t - \hat{E}(a_t|a_{t-1}, a_{t-2}, \ldots) \quad , \qquad (2.5)$$

where $\hat{E}(.)$ is the optimal linear forecast of a_t given an infinite number of past observations $(a_{t-1}, a_{t-2}, \ldots)$.

To apply the central limit theorem, it will be shown that the linearly deterministic component κ_t is zero and that the moving average coefficients are absolutely summable (which is not guaranteed by Wold's theorem, as square summability does not imply absolute summability).

First, we prove that $\kappa_t = 0$. Using the definitions of x_t and u_t, we find that

$$a_t = x_t u_t = \phi \rho x_{t-1} u_{t-1} + \rho \mu_t u_{t-1} + \phi x_{t-1} \epsilon_t + \epsilon_t \mu_t \quad .$$

Furthermore (compare Hamilton 1994, p.77),

$$\hat{E}(a_t|a_{t-1}, a_{t-2}, \ldots) = \phi \rho x_{t-1} u_{t-1} = \phi \rho a_{t-1} \quad .$$

Replacing $\hat{E}(a_t|\cdot)$ in the definition of v_t in (2.5) and using equation (2.4) yields:
$$v_t = a_t - \phi \rho a_{t-1}$$
and
$$a_t = \psi_0(a_t - \phi \rho a_{t-1}) + \psi_1(a_{t-1} - \phi \rho a_{t-2}) + \cdots + \kappa_t \quad . \qquad (2.6)$$

By definition $\psi_0 = 1$, and defining $\psi_j = \phi^j \rho^j$ implies for an infinite number of observations that equation (2.6) holds if $\kappa_t = 0$ as $\lim_{t \to \infty} \rho^t \phi^t a_0 = 0$. [8]

For a finite number of observations equation (2.6) holds approximately:

$$a_t = a_t + \phi^t \rho^t a_0 + \kappa_t \quad .$$

For t large enough $\phi^t \rho^t$ is negligible, which implies that $\kappa_t = 0$. Alternatively one could set the initial value of a_0 to zero.
Absolute summability of ψ_j can be verified easily by using its definition

---

[8] It is interesting to note that for an infinite number of observations, the product of two covariance stationary AR(1) processes also follow an AR(1) process.

through Wold's decomposition.

Now the central limit theorem (equation (2.3)) can be applied:

$$\frac{\sum_{t=1}^{T} x_t u_t}{\sqrt{T}} \overset{d}{\to} N\left(0, \sum_{j=-\infty}^{\infty} \gamma_j\right) \quad.$$

The term $\sum_{j=-\infty}^{\infty} \gamma_j$ can be calculated as follows:

$$\sum_{j=-\infty}^{\infty} \gamma_j = 2\sum_{j=0}^{\infty} \gamma_j - \gamma_0 = 2\sum_{j=0}^{\infty} \rho^j \phi^j \sigma_u^2 \sigma_x^2 - \sigma_u^2 \sigma_x^2 = \sigma_x^2 \sigma_u^2 \frac{(1+\rho\phi)}{(1-\rho\phi)} \quad.$$

The limiting distribution of t_T is calculated by applying Slutsky's theorem:

$$t_T \overset{d}{\to} (\sigma_x)^{-1}(\sigma_u)^{-1} N\left(0, \sigma_x^2 \sigma_u^2 \frac{(1+\rho\phi)}{(1-\rho\phi)}\right) = N\left(0, \frac{1+\rho\phi}{1-\rho\phi}\right) \quad. \qquad (2.7)$$

Second proof

Alternatively, the asymptotic distribution of t_T can be derived in the following way: Multiplying and dividing t_T by the true variance of b yields:

$$t_T = \frac{(x'x)^{-1}x'u}{\sqrt{\sigma_u^2(x'x)^{-1}x'\Omega_u x(x'x)^{-1}}} \sqrt{\frac{\sigma_u^2(x'x)^{-1}x'\Omega_u x(x'x)^{-1}}{s_u^2(x'x)^{-1}}} \quad. \qquad (2.8)$$

Conditional on x, the first term is exactly distributed as a $N(0,1)$. Since this distribution is independent of x, the unconditional distribution is also $N(0,1)$. The second term converges in probability: as shown in equation (2.1), s_u^2 converges to σ_u^2.

Furthermore, note that

$$\frac{x'\Omega x}{x'x} = \frac{\sum_{j=0}^{T} \frac{x_t^2}{T} + 2\rho \sum_{j=0}^{T-1} \frac{x_t x_{t-1}}{T} + 2\rho^2 \sum_{j=0}^{T-2} \frac{x_t x_{t-2}}{T} + \cdots}{\sum_{j=0}^{T} \frac{x_t^2}{T}}$$

$$\overset{p}{\to} \frac{\sigma_x^2 + 2\rho\phi\sigma_x^2 + 2\rho^2\phi^2\sigma_x^2 + \cdots}{\sigma_x^2} = \frac{1+\rho\phi}{1-\rho\phi} \quad.$$

This implies that the second term of equation (2.8) converges in probability to $\sqrt{\frac{1+\rho\phi}{1-\rho\phi}}$ (by Slutsky).

Applying Slutsky again yields:

$$t_T \overset{d}{\to} N\left(0, \frac{1+\rho\phi}{1-\rho\phi}\right) \quad.$$

2.4 Invariance results

It is important to note that the distribution of t_T is invariant with respect to σ_ϵ^2 and σ_μ^2 under the null hypothesis. As shown in equation (2.2), t_T can be written as

$$t_T = \frac{\sqrt{(T-1)}(x'u)}{\sqrt{u'ux'x - (u'x)^2}} \quad ,$$

where u_t is normally distributed. Varying the variance of u_t is equivalent to multiplying u_t by a positive constant, and will not change the value of t_T. Recall that $\sigma_u = \sigma_\epsilon^2/(1-\rho^2)$ and $\sigma_x = \sigma_\mu^2/(1-\phi^2)$. For a given ρ, σ_u^2 is proportional to σ_ϵ^2. Therefore, the small-sample distribution of t_T is unaffected if σ_ϵ^2 is varied. The same argument holds for σ_x^2 and σ_μ^2. In the simulation study, σ_ϵ^2 and σ_μ^2 are set equal to one without loss of generality.

It is also important to note that the small sample distribution of t_T under the null hypothesis is the same, e.g. for ($\rho = .1$, $\phi = .5$) and ($\rho = .5$, $\phi = .1$). This is due to the fact that x_t and u_t are both generated by an AR(1) and due to the symmetric structure of t_T formulated in equation (2.2).

Having shown the asymptotic distribution of t_T, the small sample properties of this test will be studied next. As the exact distribution is very complicated (or may even be impossible) to derive, we estimate the small sample distribution by simulation.

2.5 Monte Carlo experimentation

Monte Carlo experimentation is a method which allows one to simulate properties of a "statistic" (compare, e.g., Davidson and MacKinnon, 1993). In empirical economics, estimators or certain tests statistics are often used, for which the statistical properties are known only asymptotically. As observations on economic data are rather limited, small sample properties can be estimated by simulation.

A Monte Carlo experiment can be characterised by the objective of the study, the data-generating process (DGP) and the parameter space (compare Hendry 1984, p.940).

The DGP is the complete statistical characterisation of a model. This implies that the functional form, the distributions of the stochastic variables and the parameter values are known. In reality of course, we may just have a vague idea of the DGP. Using an ad hoc DGP will produce totally specific experimental results and cannot be generalised. One major challenge in Monte Carlo analysis is to formulate a DGP which can represent more general situations. The researcher should indicate which of his results might be specific to the construction of the experiment and why this may still be related with reality. The results may depend strongly on the parameter values chosen (see Mizon and Hendry, 1980). It is also a difficult task to justify distributional assumptions for the explanatory variables in times series, because the distributions are not observable. Finally, the results of the Monte Carlo experiment can be presented by graphics and response surfaces.

2.5.1 Simulating the distribution of the t-statistic

The relation between a homogeneity test and a t-test is straightforward in this example. The homogeneity hypothesis in demand analysis can be formulated as an omitted variable test. Here there is just one equation and one coefficient to be estimated. So, it will be tested if β differs significantly from zero. The distribution of the quasi t-statistic depends on the DGP.

The objective of this Monte Carlo experiment is to answer the following question: how fast does the small sample distribution of t_T tend to the asymptotic distribution given in equation (2.7)?

The speed of convergence can be analysed by reporting the ratio of the simulated to the asymptotic critical values of the t-test for a given nominal type I error. For increasing T, this ratio approaches one.

The left and right asymptotic critical values of the t-statistic (compare equation (2.7)) are $\pm z_{\alpha/2}\sqrt{\dfrac{1+\rho\phi}{1-\rho\phi}}$ for a given nominal type I error α, where $z_{\alpha/2}$ is the critical value from the standard normal distribution.

The power of the test is not analysed because estimation of the power function (see Mizon and Hendry 1980 for an example) is time consuming. Generalising the result for different parameter values is also quite complicated.

The DGP is defined by assumptions 2.1 and the initial values of x_t and u_t with

$$x_0 \sim N\left(0, \frac{\sigma_\mu^2}{1-\phi^2}\right) \text{ and } u_0 \sim N\left(0, \frac{\sigma_\epsilon^2}{1-\rho^2}\right).$$

The parameters are denoted by θ and T with $\theta = (\rho, \phi, \sigma_\epsilon^2, \sigma_\mu^2)' \in \Theta = \{\theta|\ |\rho| < 1, |\phi| < 1, \sigma_\epsilon^2 > 0, \sigma_\mu^2 > 0\}$ and $T \in \mathcal{T}$, where $\mathcal{T}$ is a set of positive integers with the smallest and greatest value preassigned. The parameter space is defined by $\Theta \times \mathcal{T}$. As shown before, σ_ϵ^2 and σ_μ^2 can be set equal to one without loss of generality.

The distribution of the quasi t-statistic will be characterised by its estimated critical values. Let $F_{\theta,T}$ be the distribution function of the random variable t_T for given (θ, T). The α-quantile q_α of $F_{\theta,T}$ is defined by

$$F_{\theta,T}(q_\alpha) = \alpha \quad .$$

Next, the steps of one experiment and the simulation estimators will be briefly described.

The simulation experiment

The main steps of one experiment for a given parameter vector (θ, T) are:

1) generation of x and u at every replication, [9]

2) estimation of β and calculation of the t-statistic,

3) iteration of 1) and 2) and

4) estimation of the test's lower and upper critical values for a given nominal type I error.

Simulation estimators

The parameters of interest are the critical values for a given nominal type I error and their standard errors (compare Davidson and MacKinnon 1992). The estimator of the critical value (quantile of order α) $\bar{q}_\alpha$ is defined by

$$N^{-1} \sum_{j=1}^{N} I(\bar{q}_\alpha - t_j) = \alpha \quad , \qquad (2.9)$$

where N is the number of replications and t_j is the jth value of the generated t_T values[10]. $I(\cdot)$ is an indicator function taking the value one if its argument is positive and zero otherwise. Since it cannot be assumed that the density of t_T is symmetric, lower and upper critical values are estimated: $\bar{q}_{\alpha/2}$ and $\bar{q}_{1-\alpha/2}$. In order to get a reliable result, N can be chosen rather large or alternatively, the N replications are sectioned into m groups, each consisting of n replications. The latter method is chosen because the standard error of the estimators can easily be estimated with this approach. Unlike in Davidson and MacKinnon (1992) control variates are not used [11].

The estimated critical value $\bar{q}_\alpha$ is the average of the m estimated critical values:

$$\bar{q}_\alpha := \frac{1}{m} \sum_{j=1}^{m} \bar{q}_\alpha^j \quad ,$$

where $\bar{q}_\alpha^j$ is the estimated α-quantile in the jth experiment. n is chosen such that αn is an integer and $\bar{q}_\alpha$ is calculated as the mean of the set of numbers satisfying equation (2.9) with N replaced by n and $\bar{q}_\alpha$ by $\bar{q}_\alpha^j$ [12] (compare Davidson and MacKinnon, 1992). The variance of $\bar{q}_\alpha$ is estimated by

---

[9] Here x is generated for each replication. Alternatively, one could generate the realisations of t_T conditional on x and average the results. Theory gives no answer as to which of the two methods should be preferred in general (compare Davidson and MacKinnon 1992, p.741f).

[10] The notation for the estimators is used as in Hendry (1984) with the simulation estimators denoted by "⁻".

[11] Using control variates for the estimation of quantiles for small α seems not to work well (compare Davidson and MacKinnon 1992, p.218).

[12] n must be chosen large enough to ensure that the estimation is not biased. m should be large enough such that the law of large numbers holds approximately.

$$\bar{\sigma}^2(\bar{q}_\alpha) := \frac{1}{m(m-1)} \sum_{j=1}^{m} (\bar{q}_\alpha^j - \bar{q}_\alpha)^2 \quad .$$

Finally, the quantity of interest is the ratio of the true to the asymptotic critical values, denoted by ψ and its estimator is denoted by $\bar{\psi}$.

The true type I error of the test was chosen to be $\alpha = 0.05$. Other values of α are not considered simply because the purpose of this chapter is to present an example of a Monte Carlo experimentation. The example itself is not of general interest.

The critical values are estimated for different points in the parameter space [13]

$\theta \times \mathcal{T}$ with
$\rho = \{-.95, -.75, -.55, -.35, -.15, -.05, .05, .15, .35, .55., .75, .95\}$,
$\phi = \{.15, .35, .55., .75, .95\}$,
$T = \{15, 20, 25, 30, 35, 40, 45, 50, 75, 100\}$

and $\sigma_\epsilon^2 = 1, \sigma_\mu^2 = 1$. Negative values of ϕ are not used for the simulation because of the symmetric structure of t_T (compare equation (2.2)). The special case of $\rho = 0$ or $\phi = 0$ is not considered as the exact distribution of t_T is known (student). The case of $|\rho|$ and $|\phi|$ approaching one is not of interest here, because only stationary series are considered. The number of experiments equals 600. The number of replications within one experiment is determined by the choice of m and n; here $n = 500$ and $m = 200$. In order to check if n and m are reasonable large, the estimated critical values were compared to the estimates based on equation (2.9) with $N = 100000$.

2.5.2 Presentation of the results

To make the presentation of the results easier to understand for the reader, the results are presented by response surfaces[14]. Response surfaces summarise the simulation results by regression analysis (compare Hendry 1984, or Davidson and MacKinnon 1993, p.755ff). The aim is not to find the true relation, but to find an approximation which summarises the result in a comprehensive way. The ratio of estimated small sample to asymptotic critical values is the dependent variable denoted by $\bar{\psi}$, and the independent variables are functions of the parameter values of the DGP.

The dependence between $\bar{\psi}$ and points in the parameter space can be expressed in a conditional expectations formula (compare Hendry 1984, p.949):

$$E(\bar{\psi}|\theta, T) = G(\theta, T) \quad .$$

---

[13] When doing Monte Carlo experimentation with specific economic models in mind, empirical work might give an indication of the range and/or the sign of the coefficients (yearly macrovariables in levels are usually positively correlated in time, e.g.).

[14] Tables of numerical results are not reported.

Let $\overline{\psi}_i$ be the estimated ratio in the ith experiment (the ith parameter combination). As $G(\cdot)$ is generally not known, one postulates a relation of the form

$$\overline{\psi}_i = H(\theta, T) + v_i \qquad (2.10)$$

with $v_i \sim N(0, \sigma^2(\overline{\psi}_i)), i = 1, ..., M$ where M denotes the number of experiments (here, $M = 600$). $H(\cdot)$ is an approximation of $G(\cdot)$.

The normality assumption of the errors is justified if the number of replications of one experiment is large enough. v_i is a composite error term consisting of the error from approximating $G(\cdot)$ by $H(\cdot)$ and of the experimental error. These two types of errors are assumed to be independent.

By construction of the experiment, the errors are independent (if $H(\cdot)$ is not misspecified) as the random variables were generated for each experiment independently. Furthermore, the errors are heteroskedastic as it is assumed that they depend on $\overline{\psi}_i$. By dividing the right-hand and left-hand side of equation (2.10) by the estimated standard errors $\sigma(\overline{\psi}_i)$, errors are homoskedastic. This transformation is useful because it helps to detect misspecification of the estimated response surface. As $\sigma(\overline{\psi}_i)$ is unknown it will be replaced by the estimated standard error $\overline{\sigma}(\overline{\psi}_i)$. A "good" specification requires that residuals are not heteroskedastic and that the residual variance is close to one .

The *functional form* of $H(\cdot)$ is determined as follows: it is known that for large T, $\overline{\psi}_i$ should be approximately one. Therefore, the function $H(\cdot)$ is specified to consist of a constant and terms of $o(T^c)$ with $c > 0$. Following Davidson and MacKinnon (1993, p.759ff), graphical methods are used to seek candidates of regressors for the response surfaces. As an example, two figures are shown for $\overline{\psi}_i$ related to the upper critical value (at the end of this chapter). Figure 2.1 shows the plot of the ratio of the estimated to the asymptotic upper critical value $\overline{\psi}_i$ against T for $(\rho, \phi) = (-0.95, 0.95), (0.15, 0.15), (0.95, 0.95)$ and $\alpha = 5\%$. Figure 2.2 shows the plot of $\overline{\psi}_i$ against the product of $\rho\phi$ for $T = 15, 30, 50$ and $\alpha = 5\%$. As can be seen from these figures, the ratio is large for small T and large for negative values of $\rho\phi$.

The functional form of the relationship between $\overline{\psi}$ and $\rho\phi$ in figure 2.1 can be approximated by the function $\dfrac{1}{\sqrt{T}}\sqrt{\dfrac{1-\rho\phi}{1+\rho\phi}}$. The functional form in figure 2.2 can be approximated by $\dfrac{1}{T}\dfrac{(\rho\phi)^2}{\sqrt{1+\rho\phi}}$.

Using this information, a general form of $H(\cdot)$ is chosen which is additive and consisting of a constant term and regressors divided by an appropriate positive power of T. The regressors z_1^2 and z_2^2 are chosen on the basis of the graphical analysis. The regressor z_3^2 was chosen because it helped to improve the estimation results. To be specific:

$$z_1^d := \left(\frac{1-\rho\phi}{1+\rho\phi}\right)^d \quad ,$$

$$z_2^d := \left(\frac{(\rho\phi)^2}{\sqrt{1+\rho\phi}} \right)^d$$

with $d \in \{.5, 1, 1.5, 2, 2.5, 3\}$ and

$$z_3^d := (\rho\phi)^d$$

with $d \in \{1, 2\}$.

Of course, the regressors are highly collinear and there is no unique way of reducing the number of regressors when doing estimation. This is not a severe problem in the sense that we are only interested in a good numerical approximation of $G(\cdot)$. The estimated form of $G(\cdot)$ must be able to produce reliable predictions for any point in the parameter space [15].

The estimated response function for the ratio of the estimated to the asymptotic lower critical value $\dfrac{\overline{q}_{.025}}{-1.96\sqrt{\frac{1+\rho\phi}{1-\rho\phi}}}$ is:

$$
\begin{aligned}
H(T,\rho,\phi) = \; & \underset{(.0011)}{1.0096} + \frac{1}{T}\Big(\underset{(11.9205)}{-64.2277} + \underset{(18.2003)}{101.3011}\, z_1 - \underset{(7.0244)}{40.2920}\, z_1^{1.5} + \underset{(.7476)}{4.1914}\, z_1^2 \\[4pt]
& - \underset{(11.0482)}{52.2172}\, z_2 - \underset{(2.4249)}{10.3556}\, z_2^{1.5} + \underset{(4.5560)}{38.2249}\, z_2^2 + \underset{(18.5301)}{97.1539}\, z_3 \Big) \\[4pt]
& + \frac{1}{T^{1.5}}\Big(\underset{(13.3342)}{-60.2159}\, z_1 + \underset{(15.2337)}{70.5929}\, z_1^{1.5} - \underset{(1.9484)}{8.6689}\, z_1^2 - \underset{(78.6479)}{351.9755}\, z_2 \\[4pt]
& - \underset{(21.5034)}{130.2126}\, z_2^2 + \underset{(54.6128)}{54.0400}\, z_3 + \underset{(54.6128)}{236.2589}\, z_3^2 \Big)
\end{aligned}
$$

with $s_v = 1.0555$, $c\hat{o}r(\overline{\psi}, \widehat{\overline{\psi}}) = .99765$.

The estimated response function for the ratio of the estimated to the asymptotic upper critical value $\dfrac{\overline{q}_{.975}}{1.96\sqrt{\frac{1+\rho\phi}{1-\rho\phi}}}$ is:

$$
\begin{aligned}
H(T,\rho,\phi) = \; & \underset{(.0010)}{.99306} + \frac{1}{T}\Big(\underset{(8.6833)}{59.1015} - \underset{(8.7797)}{58.8609}\, z_1 + \underset{(.1052)}{.58804}\, z_1^2 + \underset{(34.7508)}{227.2928}\, z_2 \\[4pt]
& + \underset{(4.8338)}{42.2386}\, z_2^2 - \underset{(16.9830)}{117.4214}\, z_3 - \underset{(18.4001)}{114.2710}\, z_3^2 \Big) + \frac{1}{T^{1.5}}\Big(\underset{(43.3476)}{214.2871} \\[4pt]
& - \underset{(66.2861)}{333.1231}\, z_1 + \underset{(25.6771)}{135.2052}\, z_1^{1.5} - \underset{(2.7601)}{14.3290}\, z_1^2 + \underset{(41.3589)}{185.1739}\, z_2 \\[4pt]
& - \underset{(13.0877)}{107.6577}\, z_2^2 - \underset{(67.4003)}{316.4015}\, z_3 \Big)
\end{aligned}
$$

with $s_v = 1.0331$, $c\hat{o}r(\overline{\psi}, \widehat{\overline{\psi}}) = .99781$.

---

[15] For estimating and testing the response surface regressions, the MICROFIT package (version 3.0) written by Pesaran (1991) was used.

Diagnostic tests [16] were used to test for the normality and the homoskedasticity of the residuals. For both tests the null hypothesis (normality, homoskedasticity) was not rejected. The constant terms are close to one numerically but do differ from one significantly. This is not considered as a problem. All estimates of the regression coefficients have p-values smaller than 10^{-3} when testing if they differ from zero. The estimated standard error of the residuals s_v is close to one and the estimated correlation between the simulated and fitted ψ is very high for both equations. The ability of the response surface to predict values of $\overline{\psi}_i$ (not used for the estimation of $H(\cdot)$) is checked by the predictive failure test. A sample of observations of $\overline{\psi}_i$ was sectioned into four subsamples with "smaller" values of $T \in \{17, 22, 27\}$, "larger" values of $T \in \{32, 42, 52\}$ and "smaller" values of (ρ, ϕ) with $|\rho\phi| \leq .25$, "larger" values of (ρ, ϕ) with $.25 < |\rho\phi| \leq .90$. The predictive failure test is carried out for the four subsamples individually and jointly. These tests suggest that the estimated equations can predict $\overline{\psi}_i$ well.

---

[16] These tests are implemented in the MICROFIT 3.0 package and briefly described in the user manual (appendix B).

Figure 2.1: Ratio of the estimated to the asymptotic upper critical value with T increasing for three values of (ρ, ϕ)

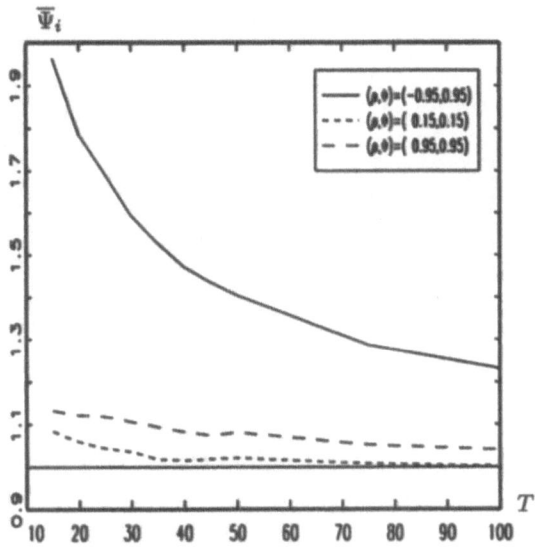

Figure 2.2: Ratio of the estimated to the asymptotic upper critical value with $\rho\phi$ increasing for three values of T

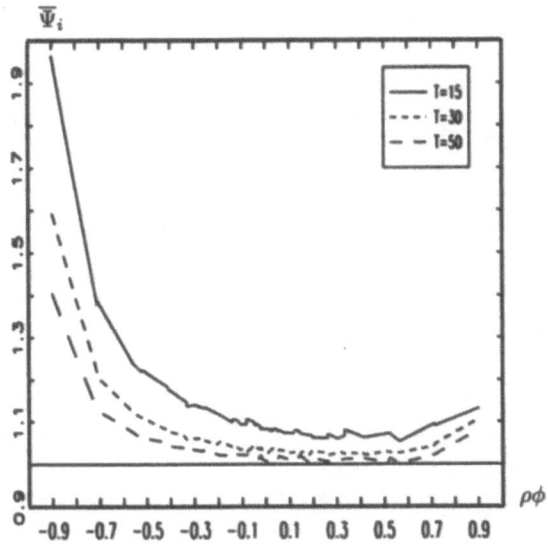

3. Consumer theory and the Rotterdam model

In this chapter, some basic concepts of consumer theory are presented (without proof) in order to describe the theoretical background of the Rotterdam model (see Barten and Böhm, 1982 and Deaton and Muellbauer, 1980). In section 3.1, we briefly define the commodity space and the budget set. The utility function and the Marshallian demand will be defined in section 3.2 and the cost function and the Hicksian demand in section 3.3. The Rotterdam model is derived section 3.4 and written in matrix form in section 3.5.

3.1 Commodity space and budget set

Let $q := (q_1, q_2, \ldots, q_r)'$ denote a commodity bundle consisting of r commodities, where q_i denotes the quantity of commodity i. The vector $q \in \mathbf{R}_+^r$ denotes a point in the commodity space which is closed, convex and bounded from below. The consumer's choice of a commodity bundle is restricted by the commodity space and the consumer's linear budget constraint

$$p'q \le e \quad ,$$

where p is a vector of positive prices $p := (p_1, p_2, \ldots, p_r)'$ with p_i being the price of the ith commodity and e (> 0) the consumer's available budget. The set within which the consumer can choose a commodity bundle is called the budget set and will be denoted by $B_{\text{SET}}(p, e)$. Formally, $B_{\text{SET}}(p, e)$ is defined by

$$B_{\text{SET}}(p, e) := \{q \in \mathbf{R}_+^r \mid p'q \le e\} \quad ,$$

and is a compact (closed and bounded) and convex set.

3.2 Preferences, direct utility function and Marshallian demand

The consumer will choose his preferred commodity bundle from the budget set. Preferences will be characterised by the preference relation $\succeq$ on the commodity space. Let $q^{(1)}$ and $q^{(2)}$ be two possible consumption bundles

$(q^{(1)} \neq q^{(1)})$. The relation $q^{(1)} \succeq q^{(2)}$ states that $q^{(1)}$ is at least as good as $q^{(2)}$. It is assumed that the preference relation satisfies the axioms of reflexivity, completeness, transitivity and continuity. These axioms and the relation $\succeq$ define a preorder which guarantees the existence of an ordinal utility function. A utility function is a numerical representation of the preference ordering and is defined by

$$U : \mathbf{R}^r_+ \mapsto \mathbf{R} : U(q) = u \quad \text{such that} \quad q^{(1)} \succeq q^{(2)} - U(q^{(1)}) \geq U(q^{(2)}) \quad .$$

Furthermore, preferences are assumed to satisfy the axioms of nonsatiation and convexity. Nonsatiation implies that the utility function is nondecreasing in q_i and increasing in at least one q_i. The consumer will, therefore, exhaust his budget. The axiom of convex preferences implies that if a consumer is indifferent between two different consumption bundles, he prefers any convex combination of the two. The economic motivation is that the consumers prefers weighted consumption bundles.

It is assumed that the utility function $U(q)$ satisfies the following properties (see Barten and Böhm 1982):

- $U(q)$ is twice continuously differentiable for all $q \in \mathbf{R}^r_+$
- $\partial U(q)/\partial q_i > 0$ for $i = 1, 2, \ldots, r$.
- $U(q)$ is strictly quasi-concave:

$U(q^{(1)}) \geq U(q^{(2)}) \Rightarrow U(\mu q^{(1)} + (1 - \mu)q^{(2)}) > U(q^{(2)})$ for $0 < \mu < 1$.
- $U(q)$ is strongly quasi-concave:

$z' \left(\frac{\partial^2 U(q)}{\partial q \partial q'} \right) z < 0 \quad \forall z \in \mathbf{R}^r$ with $(\partial U(q)/\partial q)'z = 0$ and $z \neq 0$. The derivatives are evaluated at any $q \in B_{\text{SET}}$ but for the same q.

These assumptions guarantee that the demand functions derived from the utility function are continuously differentiable.

A rational consumer will choose that bundle in the budget set which yields the highest utility level. The Marshallian (also called uncompensated) demand function results from maximising the direct utility function subject to the budget constraint. Let $q^m_i = q^m_i(p, e)$ be the Marshallian demand for commodity i and $q^m := (q^m_1, q^m_2, \ldots, q^m_r)'$. The Marshallian demand vector q^m is defined by

$$q^m(p, e) := \arg\max_q \{U(q) | q \in B_{\text{SET}}\} \quad .$$

The properties of the utility function imply that q^m is uniquely defined.

3.3 Cost function and Hicksian demand

Let $C(p, u)$ be the cost (expenditure) function which results from minimising the costs of obtaining a given utility level u. Describing the consumer's choice as a cost-minimising problem subject to a target level of utility or as a utility-maximisation problem subject to a budget constraint is the same (dual optimisation problem).

Formally, the cost function is defined by

$$C(p, u) := \min_q \{ p'q \,|\, U(q) \geq u, q \in B_{\mathrm{SET}} \} \quad .$$

The properties imposed on the utility function imply the following properties of the cost function (see Barten 1982 and Deaton and Muellbauer 1980, p.37ff):

- $C(p, u)$ is twice continuously differentiable in p,
- $C(p, u)$ is increasing in u, nondecreasing in p and increasing in at least one price.
- $C(p, u)$ is concave in p,
- $C(\mu p, u) = \mu C(p, u)$ (the cost function is homogeneous of degree 1 in p),
- $h_i(p, u) = \partial C(p, u)/\partial p_i$ for $i = 1, 2, \ldots, r$, where $h_i(p, u)$ denotes the Hicksian (compensated) demand function for q_i (Shephard's lemma).

3.4 The Rotterdam model

The Rotterdam model can be seen as a first order approximation of a demand system and is not related to a specific direct utility or cost function. For deriving the Rotterdam model, we will use the total differential of the Marshallian demand, the budget constraint, as well as the Slutsky equation (see Deaton and Muellbauer 1980, p.67ff).

The total differential of the Marshallian demand $q_i^m(p, e)$ is

$$\mathrm{d}q_i^m(p, e) = \sum_{j=1}^r \frac{\partial q_i^m(p, e)}{\partial p_j} \mathrm{d}p_j + \frac{\partial q_i^m(p, e)}{\partial e} \mathrm{d}e \quad . \tag{3.1}$$

The total differential of the budget equation $e = \sum_{i=1}^r p_j q_j$ is

$$\mathrm{d}e = \sum_{j=1}^r p_j \, \mathrm{d}q_j + \sum_{j=1}^r q_j \, \mathrm{d}p_j \quad . \tag{3.2}$$

For deriving the Slutsky equation, we will use the identity

$$h_i(p, u) \equiv q_i^m(p, C(p, u)) \quad , \tag{3.3}$$

which states that the Hicksian demand at a fixed level u is equivalent to the Marshallian demand at fixed income $e = C(p, u)$.

Taking the derivatives with respect to p_j on the left-hand and the right-hand side of the identity in (3.3) and using the rule of differentiation for composite functions yields:

$$\frac{\partial h_i(p,u)}{\partial p_j} = \frac{\partial q_i^m(p,e)}{\partial p_j} + \frac{\partial q_i^m(p,e)}{\partial e}\frac{\partial C(p,u)}{\partial p_j} \quad .$$

The term $\partial C(p,u)/\partial p_j$ will be replaced by q_j^m which follows from Shephard's lemma and the identity in (3.3). Solving then for $\partial q_i^m(p,e)/\partial p_j$ establishes the Slutsky equation:

$$\frac{\partial q_i^m(p,e)}{\partial p_j} = \frac{\partial h_i(p,u)}{\partial p_j} - \frac{\partial q_i^m(p,e)}{\partial e}q_j^m \quad . \tag{3.4}$$

Replacing the terms $\partial q_i^m(p,e)/\partial p_j$ in equation (3.1) by (3.4) and the term de in equation (3.1) by equation (3.2), and evaluating q_j at q_j^m yields:

$$\begin{aligned}
dq_i^m(p,e) &= \sum_{j=1}^{r}\left(\frac{\partial h_i(p,u)}{\partial p_j} - \frac{\partial q_i^m(p,e)}{\partial e}q_j^m\right)dp_j \\
&\quad + \frac{\partial q_i^m(p,e)}{\partial e}\left(\sum_{j=1}^{r}p_j dq_j^m(p,e) + \sum_{j=1}^{r}q_j^m dp_j\right) \\
&= \sum_{j=1}^{r}\frac{\partial h_i(p,u)}{\partial p_j}dp_j + \frac{\partial q_i^m(p,e)}{\partial e}\sum_{j=1}^{r}p_j dq_j^m(p,e) \quad . \tag{3.5}
\end{aligned}$$

Multiplying equation (3.5) by p_i/e, using the identities $dq_j^m(p,e) = q_j^m d\log q_j^m(p,e)$ and $dp_j = p_j d\log p_j$ yields:

$$\begin{aligned}
\frac{p_i q_i^m}{e}d\log q_i^m(p,e) &= \sum_{j=1}^{r}\frac{\partial h_i(p,u)}{\partial p_j}\frac{p_i p_j}{e}d\log p_j + \\
&\quad \frac{\partial q_i^m(p,e)}{\partial e}p_i\sum_{j=1}^{r}\frac{p_j q_j^m}{e}d\log q_j^m(p,e) \quad . \tag{3.6}
\end{aligned}$$

Denoting the budget share of the ith commodity by $w_i := p_i q_i^m/e$ and defining

$$\gamma_{ij} := \frac{\partial h_i(p,u)}{\partial p_j}\frac{p_i p_j}{e}$$

and

$$\beta_i := \frac{\partial q_i^m(p,e)}{\partial e}p_i$$

allows us to rewrite equation (3.6) more compactly as:

$$w_i d\log q_i^m(p,e) = \sum_{j=1}^{r}\gamma_{ij}d\log p_j + \beta_i\sum_{j=1}^{r}w_j d\log q_j^m(p,e) \quad , \tag{3.7}$$

which is the Rotterdam model.

The parameters γ_{ij} and β_i can be interpreted as follows: the parameter β_i is the marginal propensity to spend on q_i. Furthermore, note that β_i can be written alternatively as

$$\beta_i := \frac{\partial q_i^m(p,e)}{\partial e} p_i = \frac{\partial q_i^m(p,e)}{\partial e} \frac{e}{q_i^m} \frac{p_i q_i^m}{e} = \frac{\partial \log q_i^m(p,e)}{\partial \log e} w_i \quad .$$

The expression $\partial \log q_i^m(p,e)/\partial \log e$ is the expenditure elasticity of q_i which will be denoted by η_i. Therefore,

$$\beta_i = \eta_i w_i \quad ,$$

by which β_i can also be interpreted as the expenditure elasticity of q_i weighted by its budget share.

Next, the parameter γ_{ij} will be reformulated for interpretation. Using

$$\frac{p_i h_i(p,u)}{e} = \frac{p_i q_i^m(p, C(u,p))}{e} = w_i \quad ,$$

the parameter γ_{ij} can be rewritten as

$$\gamma_{ij} := \frac{\partial h_i(p,u)}{\partial p_j} \frac{p_i p_j}{e} = \frac{\partial h_i(p,u)}{\partial p_j} \frac{p_j}{h_i(p,u)} \frac{p_i h_i(p,u)}{e} = \frac{\partial \log h_i(p,u)}{\partial \log p_j} w_i \quad .$$

Note that the expression $\partial \log h_i(p,u)/\partial \log p_j$ is the compensated (net) price elasticity of q_i with respect to p_j, where compensated means that the consumer's budget will be changed such that his utility level is held constant. The compensated price elasticity will be denoted by η_{ij}^*. Therefore, γ_{ij} can be expressed as

$$\gamma_{ij} = \eta_{ij}^* w_i \quad ,$$

implying that γ_{ij} can be interpreted as the compensated price elasticity of q_i with respect to p_j weighted by its budget share. It is important to note that the uncompensated price elasticity is related to the compensated price elasticity by the Slutsky equation in elasticity form (see Deaton and Muellbauer 1980, p.62) which can be derived from the Slutsky equation in (3.4):

$$\eta_{ij} = \eta_{ij}^* - \eta_i w_j \quad ,$$

where η_{ij} is the uncompensated price elasticity with $\eta_{ij} := \partial \log q_i^m(p,e)/\partial \log p_j$. Recalling that $\beta_i = \eta_i w_i$ and $\gamma_{ij} = \eta_{ij}^* w_i$, the uncompensated price elasticity η_{ij} can be expressed as

$$\eta_{ij} = \frac{\gamma_{ij}}{w_i} - \frac{\beta_i}{w_i} w_j \quad .$$

Here we are interested in estimating the Rotterdam model with time-series data. A time index will, therefore, be added to each variable:

$$w_{it}\mathrm{d}\log q_{it}^m(p_t, e_t) = \sum_{j=1}^{r} \gamma_{ij}\mathrm{d}\log p_{jt} + \beta_i \sum_{j=1}^{r} w_{jt}\mathrm{d}\log q_{jt} \quad . \tag{3.8}$$

The variables $w_{it}\mathrm{d}\log q_{it}^m$ and $\mathrm{d}\log p_{it}$ are unobservable. They will, therefore, be approximated discretely by

$$w_{it}\mathrm{d}\log q_{it}^m \approx \overline{w}_{it}\triangle\log q_{it}^m \quad \text{and} \quad \mathrm{d}\log p_{it} \approx \triangle\log p_{it} \quad \text{for } i = 1, 2, \dots, r \quad ,$$

where $\overline{w}_{it} := (w_{it} + w_{i,t-1})/2$ and the symbol $\triangle$ denotes the annual difference operator. When we are working with annual data, we have $\triangle\log q_{it}^m = \log q_{it}^m - \log q_{i,t-1}^m$ and $\triangle\log p_{it} = \log p_{it} - \log p_{i,t-1}$. When we are working with seasonal data, we have $\triangle\log q_{it}^m = \log q_{it}^m - \log q_{i,t-4}^m$ and $\triangle\log p_{it} = \log p_{it} - \log p_{i,t-4}$.

Using the Rotterdam model as in equation (3.8), replacing the variables $w_{it}\mathrm{d}\log q_{it}$ and $\mathrm{d}\log p_{it}$ by their discrete approximations, adding a constant term θ_i and an error term u_{it}, leads to the discrete version of the statistical Rotterdam model taking the form

$$\overline{w}_{it}\triangle\log q_{it}^m(p_t, e_t) = \theta_i + \beta_i \sum_{j=1}^{r} \overline{w}_{jt}\triangle\log q_{jt}^m + \sum_{j=1}^{r} \gamma_{ij}\triangle\log p_{jt} + u_{it} \quad . \tag{3.9}$$

The constant term can be interpreted as a trendlike shift in the preferences (see Barten 1969). Note that taking first differences of a model with a time trend produces a constant.

Demand theory imposes the adding-up, homogeneity, symmetry and negativity restrictions [1]. Adding-up and homogeneity are implied by the budget constraint. Symmetry follows from the fact that the cost function $C(p, u)$ is twice differentiable and negativity follows from the concavity of the cost function (see Deaton and Muellbauer 1980, ch. 1 and 2.4).

Adding-up implies that

$$\sum_{i=1}^{r} \theta_i = 0, \quad \sum_{i=1}^{r} \beta_i = 1 \quad \text{and} \quad \sum_{i=1}^{r} \gamma_{ij} = 0 \quad \text{for } j = 1, 2, \dots, r \quad ,$$

which follows from the fact that the regressands add up to the second regressor of the Rotterdam model in equation (3.9).

Homogeneity states the absence of money illusion which is the (behavioural) hypothesis of interest here. Homogeneity implies that the Marshallian demand function is homogeneous of degree zero in prices and income and the Hicksian demand function is homogeneous of degree zero in prices. Homogeneity imposes

$$\sum_{j=1}^{r} \gamma_{ij} = 0 \quad \text{for } i = 1, 2, \dots, r. \tag{3.10}$$

---

[1] These restrictions are part of the so-called integrability conditions.

If homogeneity is satisfied for $r - 1$ equations, this is also true for the rth equation due to adding-up.

Symmetry imposes

$$\gamma_{ij} = \gamma_{ji} \quad \text{for } i, j = 1, 2, \ldots, r \text{ and } i \neq j \quad .$$

Negativity imposes that the $r \times r$ substitution matrix $\partial^2 C(p, u)/(\partial p \partial p')$ is a negative semi-definite.

For discussion of the economic interpretation of adding-up, homogeneity, symmetry, and negativity see Deaton and Muellbauer (1980b, p.15, p.43f).

The demand system consisting of r equations is singular due to adding-up and, therefore, one (arbitrarily chosen) equation will be deleted for estimation. The coefficients of the deleted equation can be estimated by using the adding-up restriction. For a more detailed discussion of the implications of adding-up, see Barten (1969), Berndt and Savin (1975) and Bewley (1986).

3.5 The Rotterdam model in matrix form

For convenience of notation, let $r := n + 1$. Stacking the demand equations for n commodities defines the incomplete demand system. For date t, the model can be written in matrix form as

$$y_t = B x_t + u_t \quad , \tag{3.11}$$

where y_t is a $n \times 1$ vector of endogenous variables, B is a $n \times (n+3)$ coefficient matrix, x_t is a $(n + 3) \times 1$ vector of regressors and u_t is a $n \times 1$ vector of errors.

The vector y_t is defined as

$$\underset{n \times 1}{y_t} = \begin{pmatrix} y_{1t} \\ y_{2t} \\ \vdots \\ y_{nt} \end{pmatrix} \tag{3.12}$$

with $y_{it} = \overline{w}_{it} \Delta \log q_{it}^m$.

The coefficient matrix B is defined as

$$\underset{n \times (n+3)}{B} = (C \,|\, S \,|\, s) = \begin{pmatrix} \theta_1 & \beta_1 & \gamma_{11} & \gamma_{12} & \cdots & \gamma_{1n} & \gamma_{1,n+1} \\ \theta_2 & \beta_2 & \gamma_{21} & \gamma_{22} & \cdots & \gamma_{2n} & \gamma_{2,n+1} \\ \vdots & \vdots & \vdots & \vdots & \vdots & \vdots & \vdots \\ \theta_n & \beta_n & \gamma_{n1} & \gamma_{n2} & \cdots & \gamma_{nn} & \gamma_{n,n+1} \end{pmatrix} , \tag{3.13}$$

where B is partitioned into the $n \times 2$ matrix C, the $n \times n$ matrix S and the $n \times 1$ vector s.

The regressor vector x_t is defined as

$$\underset{(n+3)\times 1}{x_t} = \begin{pmatrix} 1 \\ \sum_{j=1}^{n+1} \overline{w}_{jt} \Delta \log q_{jt}^m \\ \Delta \log p_{1t} \\ \Delta \log p_{2t} \\ \vdots \\ \Delta \log p_{n+1,t} \end{pmatrix} . \tag{3.14}$$

Finally, u_t is the error vector with

$$\underset{n\times 1}{u_t} = \begin{pmatrix} u_{1t} \\ u_{2t} \\ \vdots \\ u_{nt} \end{pmatrix} . \tag{3.15}$$

Next, the restriction of homogeneity (see equation (3.10)) will be written in matrix form. Using the partition of the coefficient matrix $B = (C \mid S \mid s)$, homogeneity requires

$$(S \mid s)\iota_{n+1} = 0_{n\times 1} \quad , \tag{3.16}$$

where ι_{n+1} is a $(n+1) \times 1$ column vector of ones and $0_{n\times 1}$ is a $n \times 1$ column vector of zeros.

4. Robust estimation

In chapter 3, the Rotterdam model has been derived. In this chapter, a robust method for estimating the coefficients of the Rotterdam model will be presented: the quasi-maximum likelihood estimator. The motivation for this method is that this estimator is consistent for a broad class of error processes. For statistical inferences, the variance-covariance matrix of the estimated coefficients has to be estimated. A robust estimator of the variance-covariance matrix will be discussed: the so-called heteroskedasticity and autocorrelation consistent estimator (HAC).

This chapter is organised as follows: in section 4.1, the method of quasi-maximum likelihood estimation is presented and applied to the Rotterdam model. Section 4.2 deals with the consistent estimation of the variance-covariance matrix. Two cases are considered: in section 4.2.1, it is assumed that the errors are conditionally homoskedastic and serially uncorrelated. In section 4.2.2, the so-called heteroskedasticity and autocorrelation consistent estimator (HAC) is presented. We consider the quadratic spectral kernel estimator proposed in Andrews (1991), the prewhitened quadratic spectral kernel estimator (Andrews and Monahan, 1992) and the so-called "VARHAC" estimator proposed in Den Haan and Levin (1997).

4.1 Quasi-maximum likelihood estimation

Quasi-maximum likelihood estimation consists in maximising an assumed likelihood function. This likelihood function does not necessarily correspond to the true likelihood function. The assumed likelihood function is called quasi-likelihood function. The expected scores of the quasi-likelihood function are used as the moment conditions. As these moment conditions hold for a broad class of possible error processes (Hamilton, 1994, p.431), quasi-maximum likelihood estimators are robust with respect to incorrect distributional assumptions. One advantage of this approach is that the true underlying dynamics of the error process need not be known, but under fairly general conditions the estimators are consistent and asymptotically normally distributed. The cost of this approach is that these estimators are less efficient compared with the maximum likelihood estimators where the true underlying error process is known.

Quasi-maximum likelihood estimators can be interpreted as generalised method of moments (GMM) estimators. The expected scores of the likelihood function equal zero and define the moment conditions. Here the moment conditions will be based on the assumption that the errors and explanatory variables are uncorrelated. The model equation is $y_t = Bx_t + u_t$ and the following assumption holds:

Assumption 4.1

a) $E(u_t \otimes x_\tau) = 0_{n(n+3) \times 1}$ for any t and τ.

b) $M_T := \sum_{t=1}^{T} x_t x_t'/T$ is positive definite for all $T \geq n + 3$, where M_T is a $(n + 3) \times (n + 3)$ matrix and $M_T \xrightarrow{p} M$.

Define the (possibly incorrect) log-likelihood function for observation t as

$$
\begin{aligned}
l_t(B) &= (2\pi)^{-n/2} - \frac{1}{2}\ln|\Sigma| - \frac{1}{2}\mathrm{tr}\, u_t' \Sigma^{-1} u_t \\
&= (2\pi)^{-n/2} - \frac{1}{2}\ln|\Sigma| - \frac{1}{2}\mathrm{tr}\, \Sigma^{-1}(y_t - Bx_t)(y_t - Bx_t)' \quad .(4.1)
\end{aligned}
$$

The score is defined by $\partial l_t(B)/\partial \,\mathrm{vec}\, B$. For calculating this derivative, the differential $d\, l_t(B)$ will be calculated first. The differential will be of the form $d\, l_t(B) = (\mathrm{vec}\, A)' d\,\mathrm{vec}\, B$, where $l_t(B)$ is a scalar function of the $n \times (n + 3)$ matrix B and $\mathrm{vec}\, A$ is a $n(n+3) \times 1$ vector consisting of the partial derivatives of $l_t(B)$ with respect to the elements of $\mathrm{vec}\, B$. For a detailed discussion of calculating the derivative by this technique, see Magnus and Neudecker (1988, ch. 5 and ch. 9).

The differential of $l_t(B)$ can be calculated as follows:

$$
\begin{aligned}
d\, l_t(B) &= -\frac{1}{2}\mathrm{tr}\, \left(\Sigma^{-1}(-d\, B)x_t(y_t - Bx_t)' - \Sigma^{-1}(y_t - Bx_t)x_t'(d\, B)'\right) \\
&= \mathrm{tr}\, \left(\Sigma^{-1}(d\, B)x_t(y_t - Bx_t)'\right) \quad\quad (4.2)
\end{aligned}
$$

The last equation can be rewritten by using first $\mathrm{tr}\,(A'B) = (\mathrm{vec}\, A)'\mathrm{vec}\, B$, then $\mathrm{vec}\, ABC = (C' \otimes A)\mathrm{vec}\, B$ repeatedly (see Magnus and Neudecker al. 1988, p.30, equations 4 and 5) and noting that $d\,\mathrm{vec}\, B = \mathrm{vec}\, d\, B$:

$$
\begin{aligned}
d\, l_t(B) &= (\mathrm{vec}\, \Sigma^{-1})'\mathrm{vec}\, (I_n(d\, B)x_t(y_t - Bx_t)') \\
&= (\mathrm{vec}\, \Sigma^{-1})' \left((y_t - Bx_t)x_t' \otimes I_n)\right) d\,\mathrm{vec}\, B \\
&= \left((x_t(y_t - Bx_t)' \otimes I_n)\,\mathrm{vec}\, \Sigma^{-1}\right)' d\,\mathrm{vec}\, B \\
&= \left(\mathrm{vec}\, \Sigma^{-1}(y_t - Bx_t)x_t'\right)' d\,\mathrm{vec}\, B \quad . \quad (4.3)
\end{aligned}
$$

Then, the derivative is:

$$
\frac{\partial l_t(B)}{\partial \,\mathrm{vec}\, B} = \mathrm{vec}\, \Sigma^{-1}(y_t - Bx_t)x_t' \quad . \quad (4.4)
$$

The expected value of the score equals zero as

$$E\left(\partial l_t(B)/\partial \operatorname{vec} B\right) \;=\; \operatorname{vec} E\left(\Sigma^{-1}(y_t - Bx_t)x_t'\right) = \operatorname{vec} \Sigma^{-1} E(u_t x_t')$$
$$=\; 0_{n(n+3)\times 1} \;, \tag{4.5}$$

where $E(u_t x_t') = O$ because u_t and x_t are assumed to be uncorrelated by assumption 4.1.a).

The zero expected value of the score can be interpreted as the orthogonality condition. It can be seen from equation (4.5), that the expected value of the score is independent of Σ^{-1}. The orthogonality conditions can, therefore, be simplified. Let $h(B, w_t)$ with $w_t := (y_t', x_t')'$ define a $n(n+3) \times 1$ vector-valued function:

$$h(B, w_t) = \operatorname{vec}(y_t - Bx_t)x_t' \quad . \tag{4.6}$$

The orthogonality conditions are defined by

$$Eh(B, w_t) = 0_{n(n+3)\times 1} \quad \forall t \quad .$$

The sample average of $h(\cdot)$ is given by

$$g_T(B; \mathcal{Y}_T) := \frac{1}{T} \sum_{t=1}^{T} h(B, w_t) = \operatorname{vec} \frac{1}{T} \sum_{t=1}^{T} (y_t - Bx_t)x_t' \tag{4.7}$$

with $\mathcal{Y}_T := (w_T', w_{T-1}', \ldots, w_1')'$.

As there are $n(n+3)$ orthogonality conditions and $n(n+3)$ parameters of the matrix B, setting $g_T(\cdot) = 0_{n(n+3)\times 1}$ defines uniquely the estimator $\hat{B}$:

$$0 = g_T(\hat{B}; \mathcal{Y}_T) = \operatorname{vec} \frac{1}{T} \sum_{t=1}^{T} (y_t - \hat{B}x_t)x_t' = \operatorname{vec} \frac{1}{T} \sum_{t=1}^{T} \left(y_t x_t' - \hat{B} \sum_{t=1}^{T} x_t x_t' \right) \quad . \tag{4.8}$$

The equation above implies that

$$\hat{B} = \sum_{t=1}^{T} y_t x_t' \left(\sum_{t=1}^{T} x_t x_t' \right)^{-1} \quad . \tag{4.9}$$

Note that $\hat{B}$ is a method of moments estimator which is just a special case of the GMM estimator. This quasi-maximum likelihood (or GMM) estimator is identical to the maximum likelihood estimator if the vector of disturbances is normal, homoskedastic and independent in time since the assumed likelihood function corresponds to the true likelihood function. It is also identical to the least-squares estimator.

The asymptotic properties of the quasi-maximum likelihood estimator will be analysed in the framework of GMM. In order to analyse the estimator $\hat{B}$ in the GMM framework, it is useful to use the vec form of $\hat{B}$ in equation (4.9) and to reformulate it.

Upon letting $b := \operatorname{vec} B$ and $\hat{b} := \operatorname{vec} \hat{B}$ and using the rule $\operatorname{vec} ABI = (I \otimes$

A)vec B with I being the identity matrix, the estimator in equation (4.9) can be expressed as

$$
\begin{aligned}
\hat{b} &= \operatorname{vec} \sum_{t=1}^{T} y_t x_t' \left(\sum_{t=1}^{T} x_t x_t' \right)^{-1} \\
&= \operatorname{vec} \sum_{t=1}^{T} (B x_t + u_t) x_t' \left(\sum_{t=1}^{T} x_t x_t' \right)^{-1} \\
&= \operatorname{vec} B + \operatorname{vec} \sum_{t=1}^{T} I_n u_t x_t' \left(\sum_{t=1}^{T} x_t x_t' \right)^{-1} \\
&= b + \left(\left(\sum_{t=1}^{T} x_t x_t' \right)^{-1} \otimes I_n \right) \operatorname{vec} \sum_{t=1}^{T} u_t x_t' \\
&= b + \left(\left(\sum_{t=1}^{T} x_t x_t' \right)^{-1} \otimes I_n \right) \sum_{t=1}^{T} (x_t \otimes u_t) \quad , \quad (4.10)
\end{aligned}
$$

where the last equality can be found by applying the following rule: If a, b denote some column vectors, then vec $ab' = b \otimes a$ (Magnus and Neudecker 1988, p.30, equation 3).

A sufficient set of assumptions to establish the asymptotic normality of $T^{1/2}(\hat{b} - b)$ is given in Hamilton (1994), p.414f. First, these assumptions are given in the general form as in Hamilton (1994) and then specialised to the problem discussed here.

Given the GMM estimator $\hat{b} = \operatorname{vec} \sum_{t=1}^{T} y_t x_t' (\sum_{t=1}^{T} x_t x_t')^{-1}$ and the function $g_T(B; \mathcal{Y}_T) = \operatorname{vec} \frac{1}{T} \sum_{t=1}^{T} (y_t - B x_t) x_t'$ with its argument B replaced by $b := \operatorname{vec} B$, it is assumed that:

Assumption 4.2

a) $\hat{b} \xrightarrow{p} b$, where $b := \operatorname{vec} B$ denotes the true parameter vector,

b) $T^{1/2} g_T(b; \mathcal{Y}_T) \xrightarrow{d} N(O, V)$, where V is the asymptotic variance of $T^{1/2} g_T(b; \mathcal{Y}_T)$,

c) $\partial g_T(b; \mathcal{Y}_T)/\partial b \xrightarrow{p} D'$ with the columns of D' linearly independent.

Furthermore, if assumption 4.2 holds, then

$$
T^{1/2}(\hat{b} - b) \xrightarrow{d} N(O, \Phi) \tag{4.11}
$$

with

$$
\Phi := (D V^{-1} D')^{-1} \quad . \tag{4.12}
$$

Assumption 4.2 a) holds for a broad class of processes for x_t and u_t, see White, 1984, p.47 for example. The general formulation of the assumptions 4.2 b) and c) can be made more specific to our problem.

Using the definitions of $g_T(b; y_T)$ (see equation (4.7)) and $u_t = y_t - Bx_t$, the function $g_T(b; y_T)$ can be expressed as

$$g_T(b; y_T) = \text{vec} \frac{1}{T} \sum_{t=1}^{T}(y_t - Bx_t)x_t' = \text{vec} \frac{1}{T} \sum_{t=1}^{T} u_t x_t' = \frac{1}{T} \sum_{t=1}^{T}(x_t \otimes u_t) \quad .$$

Upon letting

$$v_t := x_t \otimes u_t \quad ,$$

assumption 4.2 b) can be formulated as

$$T^{1/2}g_T(b; y_T) = T^{-1/2} \sum_{t=1}^{T} v_t \overset{d}{\to} N(O, V) \quad .$$

As noted in assumption 4.2 c), D' is the probability limit of the partial derivative of $g_T(b; y_T)$ with respect to b. Here $g_T(b; y_T)$ is a $n(n + 3) \times 1$ vector function of the $n(n + 3) \times 1$ vector b. The differential of $g_T(\cdot)$ will be of the form $A \, db$, where A is a $n(n + 3) \times n(n + 3)$ matrix consisting of the partial derivatives of $g_T(\cdot)$ with respect to b (see Magnus Neudecker 1988, chapters 5 and 9).

Taking the differential of $g_T(\cdot)$, where $g_T(\cdot) = \text{vec} \sum_{t=1}^{T}(y_t - Bx_t)x_t'/T$ and using $\text{vec} \, IBC = (C' \otimes I)\text{vec} \, B$, yields

$$d \, g_T(B; y_T) = -\text{vec} \, (d \, B) \sum_{t=1}^{T} \frac{x_t x_t'}{T} = -\left(\sum_{t=1}^{T} \frac{x_t x_t'}{T} \otimes I_n \right) d \, b \quad .$$

Since $M_T := \sum_{t=1}^{T} x_t x_t'/T$, the equation above implies that the derivative of $g_T(\cdot)$ with respect to b is

$$\frac{\partial g_T(b; y_T)}{\partial b} = -\sum_{t=1}^{T} \frac{x_t x_t'}{T} \otimes I_n = -M_T \otimes I_n \quad . \tag{4.13}$$

As

$$M_T \overset{p}{\to} M$$

by assumption 4.1 b), the derivative in equation (4.13) converges in probability to

$$\frac{\partial g_T(b; y_T)}{\partial b} \overset{p}{\to} -(M \otimes I_n) = D' = D \quad .$$

Recall that the asymptotic variance of $T^{1/2}(\hat{b} - b)$ is given by $\Phi = (DV^{-1}D')^{-1}$. As D is non-singular, Φ takes the form

$$\begin{aligned} \Phi &= \left((M \otimes I_n)V^{-1}(M \otimes I_n) \right)^{-1} \\ &= (M^{-1} \otimes I_n)V(M^{-1} \otimes I_n) \quad . \end{aligned} \tag{4.14}$$

The last equality follows from the rule that the inverse of the Kronecker product $(A \otimes B)$ is $(A^{-1} \otimes B^{-1})$ if A and B are non-singular.

Let $\hat{V}$ denote some consistent estimator of V and recall that $M_T \overset{p}{\to} M$ implying that $M_T^{-1} \overset{p}{\to} M^{-1}$. Referring to the assumption that $T^{1/2}(\hat{b} - b) \overset{p}{\to} N(O, \Phi)$, it is hoped that a good approximation of the distribution of $\hat{b}$ in finite sample is

$$\hat{b} \overset{a}{\sim} N\left(b, \hat{\Phi}/T\right) \quad , \tag{4.15}$$

with

$$\hat{\Phi} = (M_T^{-1} \otimes I_n)\hat{V}(M_T^{-1} \otimes I_n) \quad . \tag{4.16}$$

The following section deals with the consistent estimation of Φ.

4.2 Estimation of the covariance matrix of the quasi-maximum likelihood estimator

In section 4.1, the quasi-maximum likelihood estimator of the coefficients of the Rotterdam model has been derived. The motivation for this estimation method is that this estimator is consistent for a broad class of error processes. For statistical inferences, the variance-covariance matrix of the estimated coefficients has to be estimated. In what follows, the consistent estimation of the variance-covariance matrix is considered for two cases: in section 4.2.1, it is assumed that the errors are homoskedastic and serially uncorrelated. In section 4.2.2, the so-called heteroskedasticity and autocorrelation consistent estimator (HAC) is presented. We consider the spectral kernel estimator proposed in Andrews (1991), the prewhitened spectral kernel estimator (Andrews and Monahan, 1992) and the so-called "VARHAC" estimator proposed in Den Haan and Levin (1997).

In the previous section, the asymptotic variance-covariance matrix of $T^{1/2}(\hat{b} - b)$, where $\hat{b}$ is the GMM estimator, was given by $\Phi = (M^{-1} \otimes I_n)V(M^{-1} \otimes I_n)$. The consistent estimation of Φ requires the consistent estimation of V, where V denotes the asymptotic variance of $T^{1/2}g_T(b; y_T)$. Recall that $T^{1/2}g_T(b; y_T) = T^{-1/2} \sum_{t=1}^{T} v_t$, where $v_t := x_t \otimes u_t$.

Throughout this chapter, we make the following assumptions:

Assumption 4.3

a) $\hat{B} \overset{p}{\to} B$

b) $V := \lim_{T \to \infty} Var(T^{-1/2} \sum_{t=1}^{T} v_t)$, which is a common assumption in time-series analysis. A set of sufficient conditions imposed on the sequence of v_t which guarantees that this assumption holds is given in White (1982, ch. 5, p.124ff), e.g.

c) $M_T := \sum_{t=1}^{T} x_t x_t'/T \overset{p}{\to} M$ and $\lim_{T \to \infty} E(\sum_{t=1}^{T} x_t x_t'/T) = M$.

In order to define a consistent estimator of V, two cases are considered. In the first case, it is assumed that u_t is homoskedastic and serially uncorrelated. In the second case, a restricted form of a time-dependent heteroskedastic process for v_t is allowed.

4.2.1 Estimation of Φ if the errors are homoskedastic and serially uncorrelated

In this subsection, it will be assumed that the errors are homoskedastic and serially uncorrelated. The following assumptions will be used to derive a consistent estimator of Φ.

Assumption 4.4

a) $E(u_t u'_{t-j}) = \begin{cases} \Sigma & \text{if } j = 0 \\ O & \text{if otherwise} \end{cases}$

b) $\sum_{t=1}^{T} u_t u'_t / T \overset{p}{\to} \Sigma$

Proposition 4.1

If assumptions 4.3 and 4.4 hold, a consistent estimator of Φ is given by

$$\hat{\Phi} := M_T^{-1} \otimes \hat{\Sigma}$$

with $\hat{\Sigma} := \sum_{t=1}^{T} \hat{u}_t \hat{u}'_t / T$ and $\hat{u}_t := y_t - \hat{B} x_t$.

Proof

Using assumption 4.4, the expression for V can be simplified such that a consistent estimator of V can be found directly. Using the fact that $E(v_t) = 0$, where $v_t := x_t \otimes u_t$, the asymptotic variance V can be reformulated as

$$V := \lim_{T \to \infty} Var(T^{-1/2} \sum_{t=1}^{T} v_t) = \lim_{T \to \infty} \frac{1}{T} E(\sum_{t=1}^{T} v_t \sum_{t=1}^{T} v'_t) = \lim_{T \to \infty} E(\sum_{t,\tau=1}^{T} \frac{v_t v'_\tau}{T}).$$

Using the definition of $v_t := x_t \otimes u_t$, the term $v_t v'_\tau$ can be reformulated as

$$v_t v'_\tau = (x_t \otimes u_t) \cdot (x_\tau \otimes u_\tau)' = (x_t x'_\tau \otimes u_t u'_\tau) \quad .$$

The expectation of $v_t v'_\tau$ is given by

$$E(v_t v'_\tau) = E(x_t x'_\tau \otimes u_t u'_\tau) = E(x_t x'_\tau) \otimes E(u_t u'_\tau) \quad ,$$

where the last equality can be found by applying the law of iterated expectations.

Using this result and the fact that u_t is serially uncorrelated, it follows that

$$E(v_t v'_\tau) = \begin{cases} E(x_t x'_\tau) \otimes E(u_t u'_\tau) & \text{for } t = \tau \\ O & \text{for } t \neq \tau \end{cases}$$

Therefore, V can be expressed as

$$
\begin{aligned}
V &= \lim_{T\to\infty} \frac{1}{T} \sum_{t,\tau=1}^{T} E(v_t v'_\tau) = \lim_{T\to\infty} \frac{1}{T} \sum_{t=1}^{T} E(v_t v'_t) \\
&= \lim_{T\to\infty} \frac{1}{T} \sum_{t=1}^{T} (E(x_t x'_t) \otimes E(u_t u'_t)) \\
&= \lim_{T\to\infty} E \sum_{t=1}^{T} \frac{x_t x'_t}{T} \otimes \Sigma \overset{p}{\to} M \otimes \Sigma \quad ,
\end{aligned}
\tag{4.17}
$$

where the last step follows from the assumption that $\lim_{T\to\infty} E \sum_{t=1}^{T} x_t x'_t / T = M$.

Next an estimator of V will be defined and, moreover, it will be shown that this estimator is consistent.

Let

$$
\hat{V} := (M_T \otimes \hat{\Sigma}) \quad ,
$$

where

$$
\hat{\Sigma} := \frac{1}{T} \sum_{t=1}^{T} \hat{u}_t \hat{u}'_t
$$

with

$$
\hat{u}_t := y_t - \hat{B} x_t \quad .
$$

The consistency of $\hat{V}$ requires the consistency of M_T and $\hat{\Sigma}$.

By assumption $M_T \overset{p}{\to} M$. The consistency of $\hat{\Sigma}$ can be shown as follows. (The proof is the multivariate analogue of that in Hamilton, p.211). It will be proven that $\sum_{t=1}^{T} \hat{u}_t \hat{u}'_t / T$ has the same probability limit as that of $\sum_{t=1}^{T} u_t u'_t / T$ which is Σ. Note that

$$
\begin{aligned}
\sum_{t=1}^{T} \frac{u_t u'_t}{T} &= \frac{1}{T} \sum_{t=1}^{T} (y_t - B x_t)(y_t - B x_t)' \\
&= \frac{1}{T} \sum_{t=1}^{T} \left(y_t - \hat{B} x_t + \hat{B} x_t - B x_t \right)(y_t - \hat{B} x_t + \hat{B} x_t - B x_t)' \\
&= \frac{1}{T} \sum_{t=1}^{T} (y_t - \hat{B} x_t)(y_t - \hat{B} x_t)' + \frac{1}{T} \sum_{t=1}^{T} (\hat{B} x_t - B x_t)(\hat{B} x_t - B x_t)' + \\
&\quad \frac{1}{T} \sum_{t=1}^{T} (y_t - \hat{B} x_t)(\hat{B} x_t - B x_t)' + \frac{1}{T} \sum_{t=1}^{T} (\hat{B} x_t - B x_t)(y_t - \hat{B} x_t)' \\
&= \sum_{t=1}^{T} \frac{\hat{u}_t \hat{u}'_t}{T} + (\hat{B} - B) \sum_{t=1}^{T} \frac{x_t x'_t}{T} (\hat{B} - B)' + \sum_{t=1}^{T} \frac{\hat{u}_t x'_t}{T} (\hat{B} - B)' +
\end{aligned}
$$

$$(\hat{B} - B) \sum_{t=1}^{T} \frac{x_t \hat{u}_t'}{T} \tag{4.18}$$

The second term of the last line in equation (4.18) converges in probability to O as $\hat{B} \xrightarrow{P} B$ and $\sum_{t=1}^{T} x_t x_t' / T \xrightarrow{P} M$ with M being a finite matrix. The third and fourth terms of equation (4.18) equal zero due to the sample orthogonality conditions. It follows that

$$\sum_{t=1}^{T} \frac{\hat{u}_t \hat{u}_t'}{T} - \sum_{t=1}^{T} \frac{u_t u_t'}{T} \xrightarrow{P} O \ .$$

Having shown that $\sum_{t=1}^{T} \hat{u}_t \hat{u}_t' / T - \sum_{t=1}^{T} u_t u_t' / T \xrightarrow{P} O$ and noting that $\sum_{t=1}^{T} u_t u_t' / T \xrightarrow{P} \Sigma$, it follows directly that

$$\hat{\Sigma} \xrightarrow{P} \Sigma$$

and, therefore,

$$\hat{V} = M_T \otimes \hat{\Sigma} \xrightarrow{P} M \otimes \Sigma \ .$$

Recalling that Φ denoted the asymptotic variance of $T^{1/2}(\hat{b} - b)$ with

$$\Phi = (M^{-1} \otimes I_n) V (M^{-1} \otimes I_n) \ ,$$

a consistent estimator of Φ can be found by estimating M^{-1} by M_T^{-1} and $\hat{V} := M_T^{-1} \otimes \hat{\Sigma}$:

$$\begin{aligned} \hat{\Phi} &= (M_T^{-1} \otimes I_n)(M_T \otimes \hat{\Sigma})(M_T^{-1} \otimes I_n) \\ &= M_T^{-1} \otimes \hat{\Sigma} \end{aligned} \tag{4.19}$$

The estimated variance of $\hat{b}$ is simply $\hat{\Phi}/T$.

4.2.2 Estimation of Φ if the errors are dependent and heterogeneous

Assume that v_t is a sequence of dependent heterogeneously distributed observations. Consistent estimators of V are called heteroskedasticity and autocorrelation consistent (HAC) estimators. Here two types of HAC estimators are considered. One class of HAC estimators based on non-parametric kernel estimation is discussed in Andrews (1991) and Andrews and Monahan (1992). The other type of HAC estimator is the parametric estimator proposed in Den Haan and Levin (1997). The HAC estimator of Φ takes the form

$$\hat{\Phi}_{HAC} := (M_T^{-1} \otimes I_n) \hat{V}_{HAC} (M_T^{-1} \otimes I_n) \ ,$$

where $\hat{V}_{HAC}$ is the HAC estimator of V.

The quadratic spectral kernel estimator

In this subsection the basic ideas of the quadratic spectral (QS) kernel estimator are described. Recall that it was assumed that the asymptotic variance V is given by

$$V := \lim_{T \to \infty} Var(T^{-1/2} \sum_{t=1}^{T} v_t)$$

which is equal to the definition of the asymptotic variance of the sample mean of a mean zero covariance stationary sequence with absolutely summable autocovariances (see Hamilton, 1994 p.280). Spectral based estimators of V are numerically equal to 2π times an estimator of the spectral density of v_t at frequency zero. A spectral estimator of V takes the form of

$$\hat{V}_{HAC} := \hat{\Gamma}_0 + \sum_{j=1}^{m} k(j, m)(\hat{\Gamma}_j + \hat{\Gamma}'_j)$$

where $\hat{\Gamma}_j$ denotes the jth estimated autocovariance matrix defined by

$$\hat{\Gamma}_j := \frac{1}{T} \sum_{t=j+1}^{T} \hat{v}_t \hat{v}'_{t-j} \tag{4.20}$$

with $\hat{v}_t := x_t \otimes \hat{u}_t$ and $\hat{u}_t := y_t - \hat{B}x_t$. The kernel $k(j, m)$ is a weighting function defined on a certain interval which is the bandwidth for smoothing the autocovariances. Note that the estimator $\hat{V}$ defined above with $k(j, m) = 1$ for all m, is proposed in White (1984, p.152) but has the disadvantage that the estimated variance-covariance matrix is not necessarily positive semi-definite in finite sample. As noted in Newey and West (1987), positive semi-definiteness can be imposed by choosing an appropriate kernel $k(\cdot)$. Andrews (1991) showed that the quadratic spectral (QS) kernel estimator has the smallest asymptotic mean squared error (MSE) within the class of kernel estimators which generate positive semi-definite estimators in finite sample.

The QS estimator for V is defined by

$$\hat{V}_{QS} := \frac{T}{T - (n + 3)} \left[\hat{\Gamma}_0 + \sum_{j=1}^{T-1} k_{QS}(j/\hat{S}_T)(\hat{\Gamma}_j + \hat{\Gamma}'_j) \right] , \tag{4.21}$$

where $T/(T - (n + 3))$ is a small sample correction factor, $k_{QS}(a)$ is the quadratic spectral kernel with $a = j/\hat{S}_T$, which is defined by

$$k_{QS}(a) := \frac{25}{12\pi^2 a^2} \left(\frac{\sin(6\pi a/5)}{6\pi a/5} - \cos(6\pi a/5) \right)$$

and $\hat{S}_T$ is a data-dependent bandwidth parameter. The bandwidth parameter used will be a function of the data and is a parametric estimate of the asymptotically optimal bandwidth. The optimal bandwidth for the quadratic spectral kernel (Andrews and Monahan 1991, equation (5.3)) is estimated by

$$\hat{S}_T = 1.3221[\hat{\alpha}T]^{1/5} \quad . \tag{4.22}$$

The form of the estimator $\hat{\alpha}$ depends on which parametric model is assumed to approximate $\hat{v}_t$ well. Andrews (1991) proposes to use AR(1) models for each component of $\hat{v}_t$. In our application, $n(n+3)$ equations have to be estimated since $\hat{v}_t$ is a $n(n+3) \times 1$ vector. Apart from parsimony, this procedure facilitates the estimation of the bandwidth parameter (see Andrews 1991, p.833-836). Using this approach $\hat{\alpha}$ is given by

$$\hat{\alpha} = \frac{\sum\limits_{j=1}^{n(n+3)} \omega_j \dfrac{4\hat{\rho}_j^2 \hat{\sigma}_j^4}{(1-\hat{\rho}_j)^8}}{\sum\limits_{j=1}^{n(n+3)} \omega_j \dfrac{\hat{\sigma}_j^4}{(1-\hat{\rho}_j)^4}} \quad , \tag{4.23}$$

where $\hat{\rho}_j$ denotes the estimated coefficient of the AR(1) model without constant for the jth component of $\hat{v}_t$, the estimated residual variance is denoted by $\hat{\sigma}_j^2$ and ω_j denotes the weight given to these estimates. Andrews (1991) proposes to give zero weight to coefficients related to the constant term (these are the first n elements of $\hat{v}_t$) and unit weight to the others. Alternative procedures are also described in Andrews (1991).

Andrews (1991) gives sufficient conditions which guarantee the consistency of the QS (and other kernel) estimator. Weaker conditions as those in Andrews (1991) are given e.g. in Hansen (1992). These conditions are very technical and will not be discussed here.

According to Den Haan and Levin (1997), setting $\omega_j = 1$ for all j has some disadvantages in small samples. If one variable is rescaled, the values of the related $\hat{\rho}_j$ are unaltered but not the value of the $\hat{\sigma}_j$ and by this the estimated bandwidth parameter changes.

Another problem is that *one* bandwidth parameter has to be used for estimating the variances of *several* estimated parameters. Den Haan and Levin (1997) propose to give weight only to components of $\hat{v}_t$ which are related to the parameters of interest. The idea behind this approach is that if one is only interested in estimating the variance of a specific estimated parameter, the bandwidth parameter should be chosen such that the small sample MSE of this estimated variance is minimised. Simulation results in Den Haan and Levin (1997) illustrate this problem.

Prewhitened quadratic spectral kernel estimator

Instead of estimating the spectrum for $\hat{v}_t$, Andrews and Monahan (1992) propose a prewhitened kernel HAC estimator. This estimator will be called

PW-QS estimator. The basic idea of this approach is that $\hat{v}_t$ will be filtered. Let the filtered series be denoted by $\hat{v}_t^*$ for which the spectrum at frequency zero will be estimated for $\hat{v}_t^*$. The motivation for this approach is that $\hat{v}_t^*$ has a flatter spectrum and, therefore, kernel estimation is likely to be less biased (see Andrews and Monahan 1992, p.703).

In order to implement the filtering procedure, Andrews and Monahan (1992) propose to estimate a VAR(d) model for $\hat{v}_t$:

$$\hat{v}_t = \sum_{r=1}^{d} \hat{A}_r \hat{v}_{t-r} + \hat{v}_t^* \qquad \text{for } t = d+1, d+2, \ldots, T \quad ,$$

where $\hat{A}_r$ is a $n(n+3) \times n(n+3)$ matrix of estimated parameters. The vector $\hat{v}_t^*$ is a $n(n+3) \times 1$ residual vector which can be interpreted as the filtered series of $\hat{v}_t$. As noted in Andrews and Monahan (1992), a VAR model should not be considered as the true model. The VAR model is used to weaken some of the temporal dependence in $\hat{v}_t$ and, therefore, the residuals of $\hat{v}_t^*$ are closer to white noise than $\hat{v}_t$, implying that the spectrum of $\hat{v}_t^*$ is flatter. Andrews and Monahan (1992) propose to use a VAR(1) model in the prewhitening procedure.

Let $\hat{V}^*$ be the estimated variance-covariance matrix of $\hat{v}_t^*$, where $\hat{V}^*$ is defined by replacing $\hat{v}_t$ with $\hat{v}_t^*$ in equation (4.20) and applying equation (4.21). Furthermore let $\hat{V}_{PW}$ denote the PW-QS estimator of V. Using the one-to-one relation between the spectrum of $\hat{v}_t$ and $\hat{v}_t^*$ (see Hamilton 1994, p.267), $\hat{V}_{PW}$ is defined by

$$\hat{V}_{PW} := (I_{n(n+3)} - \sum_{r=1}^{d} \hat{A}_r)^{-1} \hat{V}^* (I_{n(n+3)} - \sum_{r=1}^{d} \hat{A}_r')^{-1} \quad , \qquad (4.24)$$

where $I_{n(n+3)}$ is a $n(n+3) \times n(n+3)$ identity matrix.

Finally, it should be noted that the sequence of $\hat{v}_t$ may be filtered by procedures other than the one above. In fact, simulation results in Den Haan and Levin (1997) illustrate that the choice of the prewhitening procedure can have a substantial impact on the results in small samples. Their results imply that using different filtering procedures may lead to conflicting inference in small samples.

The "VARHAC" estimator

A parametric HAC estimator is proposed by Den Haan and Levin (1997). The basic idea of the so-called VARHAC estimator is to estimate a (restricted) VAR representation for $\hat{v}_t$. This estimated VAR representation is interpreted as the true VAR representation. The asymptotic variance of the sample mean of a covariance-stationary VAR process is known; it is identical to the spectrum of the VAR process evaluated at frequency zero.

Without going into details, it is interesting to note that the (asymptotic) trade-off between the bias and variance for the QS estimator depends on the size of the bandwidth (the larger the bandwidth, the smaller the bias and larger the variance) and for the VARHAC estimator, the trade-off depends on the lag order of the VAR model (the more lags, the smaller the bias and larger the variance).

The estimated VAR representation is based on some model selection criterion. The residuals resulting from this estimated VAR model are white noise for T large enough under quite general conditions.

As above, the estimated VAR(d) model for $\hat{v}_t$ takes the form

$$\hat{v}_t = \sum_{r=1}^{d} \hat{A}_r \hat{v}_{t-r} + \hat{v}_t^* \qquad \text{for } t = d+1, d+2, \ldots, T \quad ,$$

where $\hat{A}_r$ is a $n(n+3) \times n(n+3)$ matrix of estimated parameters and $\hat{v}_t^*$ is a $n(n+3) \times 1$ residual vector.

Den Haan and Levin (1997) propose to estimate each equation of the VAR model by ordinary least-squares. This means that each component of $\hat{v}_t$ is regressed on its own lagged value and on the lagged values of all other components of $\hat{v}_t$. In order to decide if some coefficients of the estimated equations should be restricted to zero, they suggest to choose the "appropriate" model by the Schwarz Bayesian criterion (BIC) or the Akaike information criterion (AIC). Let $\hat{v}_{it}$ denote the ith component of $\hat{v}_t$ and f be the number of freely estimated parameters.

The BIC criterion for the ith equation is defined by

$$BIC(d, i) := \log \left(\sum_{t=d+1}^{T} \frac{\hat{v}_{it}^2}{T} \right) + \left(\frac{\log T}{T} f \right)$$

and the AIC for the ith equation is defined by

$$AIC(d, i) := \log \left(\sum_{t=d+1}^{T} \frac{\hat{v}_{it}^2}{T} \right) + \left(\frac{2f}{T} \right) \quad .$$

The first term of the BIC and AIC criterion measures the fit of the model and the second term is a penalty function which depends on the number of freely estimated parameters. The chosen model is a compromise between the fit of the model and the number of parameters used. The BIC criterion chooses more parsimonious models compared to the AIC criterion because increasing the number of regressors is penalised more heavily. Having chosen which selection criterion to use, the preferred regression equation for $\hat{v}_{it}$ is the one with the highest value of the criterion value.

Once the preferred regression equations for the $\hat{v}_{it}$'s have been found, the matrices $\hat{A}_r$ of the VAR model for $\hat{v}_t$ can be constructed.

The covariance matrix of $\hat{v}_t^*$ is estimated by

$$\hat{V}^* := \sum_{t=d+1}^{T} \frac{\hat{v}_t^* \hat{v}_t^{*\prime}}{T - (n+3)} \quad ,$$

which can be interpreted as the spectral estimator of the asymptotic variance of the sample mean of a white noise process. Using the denominator $T - (n+3)$ instead of T as in Den Haan and Levin (1997) should be seen as a small sample degrees of freedom correction. The estimated variance-covariance matrix of $\hat{v}_t$, can be found by using the one-to-one relation between the spectrum of $\hat{v}_t$ and $\hat{v}_t^*$ which is given by

$$\hat{V}_{VH} := (I_{n(n+3)} - \sum_{r=1}^{d} \hat{A}_r)^{-1} \hat{V}^* (I_{n(n+3)} - \sum_{r=1}^{d} \hat{A}_r')^{-1} \quad ,$$

where $I_{n(n+3)}$ is a $n(n+3) \times n(n+3)$ identity matrix.

The set of sufficient conditions guaranteeing the consistency of the VARHAC estimator are similar to the ones given in Andrews and Monahan (1992). A detailed discussion of these assumptions and the convergence rate of these estimators are presented in Den Haan and Levin (1997). The main result is that the VARHAC estimator converges in mean square at least as fast as the QS and PW-QS estimator under similar conditions as outlined in Andrews (1991) and Andrews and Monahan (1992).

Comparing the VARHAC estimator $\hat{V}_{VH}$ and PW-QS estimator $\hat{V}_{PW}$, one can see that both estimators "filter" the residual vector $\hat{v}_t$. The procedure of Den Haan and Levin (1997) consists in estimating a restricted VAR model for $\hat{v}_t$ based on the BIC or AIC selection criterion which could also be used in the prewhitening procedure. In this sense, the VARHAC estimator could be seen as a PW-QS estimator which gives unit weight to the variance of the filtered residuals $\hat{v}_t^*$ and zero weight to the autocovariances of higher order.

As mentioned in Den Haan and Levin (1997), the QS estimator is based solely on an asymptotic criterion. This is also true for the PW-QS estimator if filtering is done mechanically. Using a selection criterion introduces a small sample criterion. As an analogy one can think of the GMM estimator derived in section 4.1, which is a consistent estimator in general and which simultaneously satisfies the small sample criterion of being a least-squares estimator (for the special case considered there).

Den Haan and Levin (1997) compare the QS/PW-QS and VARHAC estimator by Monte Carlo simulation. As in Andrews and Monahan (1992), they estimate coverage probabilities of the quasi t-test for univariate regression models. The superiority of one of the estimators depends on the design of the experiment. Their experiments, nevertheless, show several things worth mentioning. The performance of the QS and PW-QS estimator depends on the estimation procedure of the bandwidth. Furthermore, the choice of the prewhitening procedure can have a substantial effect on the performance of the PW-QS estimator. The user must, therefore, choose some appropriate weighting scheme and a filtering procedure for the PW-QS estimator. In this sense, kernel estimation with a data dependent bandwidth should not be seen as a mechanical exercise.

5. Testing for homogeneity

In this chapter, various tests for homogeneity will be discussed. In section 5.1, an exact test for the case of normal, homoskedastic errors which are independent in time is derived. This test corresponds to Anderson's $\mathcal{U}$ test (see Anderson 1984, 298ff). In section 5.2, it will be shown that Anderson's $\mathcal{U}$ statistic is functionally equivalent to Laitinen's statistic (see Laitinen 1978). In section 5.3, it is assumed that the errors follow a vector autoregressive process. The likelihood ratio statistic for testing homogeneity will be defined. Since only the asymptotic distribution of this test statistic is known, a small sample version of the likelihood ratio test and a Monte Carlo test is proposed. In section 5.4, we study the consequences of testing for homogeneity if the error process is wrongly specified. It will be shown that Anderson's $\mathcal{U}$ statistic derived in section 5.1 is asymptotically equivalent to a quadratic form in normal variables under quite general assumptions. In section 5.5, a robust Wald test is defined. This Wald test is based on the quasi-maximum likelihood estimation of the coefficients of the Rotterdam model and on the heteroskedasticity and autocorrelation consistent estimation of its variance-covariance matrix. This test is robust in the sense that the Wald statistic is asymptotically distributed as a χ^2 under the null hypothesis of homogeneity under fairly general assumptions.

5.1 Anderson's $\mathcal{U}$ test if the errors are time-independent (LRU)

In this section, a test for homogeneity is considered for the case, where the errors of the model are assumed to be normal, homoskedastic and independent in time.

The model is defined by $y_t = Bx_t + u_t$ and the following assumption holds:

Assumption 5.1

The $n \times 1$ error vector u_t is distributed as

$$u_t \sim i.i.d.N(O, \Sigma) \qquad \forall t \quad .$$

This regression model can be seen as a seemingly unrelated regression model (SUR) with identical regressors in each equation.

Here we are interested in testing homogeneity. Recall the partition of the $n \times (n+3)$ coefficient matrix $B = (C \mid S \mid s)$, where C is of dimension $n \times 2$, S is $n \times n$ and s is $n \times 1$ (see equation (3.13)). Homogeneity restricts the coefficient submatrix $(S \mid s)$ by (see equation (3.16))

$$(S \mid s)\iota_{n+1} = 0_{n \times 1} \quad , \tag{5.1}$$

where ι_{n+1} denotes a $(n+1) \times 1$ vector of ones and $0_{n \times 1}$ a $n \times 1$ vector of zeros.

The system of equations $y_t = Bx_t + u_t$ will be rewritten in an equivalent form such that testing for homogeneity can be interpreted as a test for the omission of a variable. The purpose of doing so is that testing for the omission of a variable in a multivariate regression model, where the errors are normal, homoskedastic and independent in time, is a known problem and treated in detail in Anderson (1984, chapter 8).

Let the $(n+3) \times 1$ regressor vector x_t be partitioned such that

$$x_t = \begin{pmatrix} x_{1t} \\ x_{2t} \\ x_{3t} \end{pmatrix} \quad ,$$

where x_{1t} is a 2×1 vector, x_{2t} is a $n \times 1$ vector and x_{3t} is a scalar. Using this partition of x_t and the partition of $B = (C \mid S \mid s)$, the model equation can be written as:

$$y_t = Bx_t + u_t = (C \mid S \mid s) \begin{pmatrix} x_{1t} \\ x_{2t} \\ x_{3t} \end{pmatrix} + u_t = Cx_{1t} + Sx_{2t} + sx_{3t} + u_t \quad .$$

Adding $S\iota_n$ to and subtracting $S\iota_n$ from the last column of the coefficient matrix B (which is s) and noting that $(s + S\iota_n) = (S \mid s)\iota_{n+1}$, the equation system can rewritten as

$$
\begin{aligned}
y_t &= Cx_{1t} + Sx_{2t} + (s + S\iota_n - S\iota_n)x_{3t} + u_t \\
&= Cx_{1t} + S(x_{2t} - \iota_n x_{3t}) + (S \mid s)\iota_{n+1}x_{3t} + u_t \\
&= (C \mid S \mid s^*) \begin{pmatrix} x_{1t} \\ x_{2t} - \iota_n x_{3t} \\ x_{3t} \end{pmatrix} + u_t \quad ,
\end{aligned}
\tag{5.2}
$$

where $s^* := (S \mid s)\iota_{n+1}$. Equation (5.2) can be written more compactly as

$$y_t = B^* x_t^* + u_t \quad , \tag{5.3}$$

where the definitions of B^* and x_t^* follow directly from equation (5.2). The last column of the transformed coefficient matrix B^*, which was denoted by s^*, equals zero under homogeneity. This follows from the definition of s^* and the homogeneity restriction in equation (5.1):

$$s^* := (S \mid s)\iota_{n+1} = 0_{n \times 1} \quad .$$

The hypotheses for testing homogeneity can be formulated as

$$H_0 : \quad s^* = 0_{n \times 1} \qquad \text{and} \qquad H_1 : \quad s^* \neq 0_{n \times 1} \quad .$$

This result is important since testing for homogeneity in a demand system can be interpreted as a test for the omission of a variable in a multivariate regression model. If the errors are normal, homoskedastic and independent in time, the exact distribution of a monotonic transformation of the likelihood ratio is given in Anderson (1984, p.298ff). It turns out that this is the same statistic as proposed in Laitinen (1978) which is a multiple of Hotelling's $\mathcal{T}^2$ statistic.

The likelihood ratio λ is defined by

$$\lambda := \frac{\max\limits_{B^*, \Sigma \in H_0} L(B^*, \Sigma)}{\max\limits_{B^*, \Sigma \in H_0 \cup H_1} L(B^*, \Sigma)} \quad , \tag{5.4}$$

where $L(B^*, \Sigma)$ is the likelihood function related to the equation $y_t = B^* x_t^* + u_t$ for $t = 1, 2 \ldots, T$ with u_t being normal, homoskedastic and independent in time. The log-likelihood function takes the form

$$
\begin{aligned}
\ln L(B^*, \Sigma) &= -\frac{nT}{2} \ln 2\pi - \frac{T}{2} \ln |\Sigma| - \sum_{t=1}^{T} \left(\frac{1}{2} \operatorname{tr} u_t' \Sigma^{-1} u_t \right) \\
&= -\frac{nT}{2} \ln 2\pi - \frac{T}{2} \ln |\Sigma| - \\
&\quad \sum_{t=1}^{T} \frac{1}{2} \operatorname{tr} \Sigma^{-1} (y_t - B^* x_t^*)(y_t - B^* x_t^*)'.
\end{aligned}
\tag{5.5}
$$

Note that this log-likelihood function has the same form as the quasi-likelihood function discussed in section 4.1. Following Anderson (1984, ch.8), the likelihood ratio can be formulated as

$$\lambda := \frac{|\check{\Sigma}|^{-T/2}}{|\hat{\Sigma}|^{-T/2}} \quad ,$$

where $\hat{\Sigma}$ denotes the unrestricted and $\check{\Sigma}$ the restricted maximum likelihood estimator of the residual variance-covariance matrix. The estimators are given by

$$\hat{\Sigma} := \sum_{t=1}^{T} \frac{\hat{u}_t \hat{u}_t'}{T}$$

with $\hat{u} := y_t - \hat{B}^* x_t^*$, where $\hat{B}^*$ is the unrestricted estimator of B^* and

$$\check{\Sigma} := \sum_{t=1}^{T} \frac{\check{u}_t \check{u}_t'}{T}$$

with $\check{u} := y_t - \check{B}^* x_t^*$, where $\check{B}^*$ is the restricted estimator of B^*.

Here the likelihood function was defined for the transformed model equation to emphasise that the homogeneity restriction can be represented as a restriction on one column of the coefficient matrix B^*. It should, however, be noted that the restricted and unrestricted residuals resulting from the estimation of the untransformed model $y_t = B x_t + u_t$ will take exactly the same values as those of the transformed model. This follows from the fact that the underlying probability distribution of the model is the same for the transformed and untransformed model. This implies that

$$\hat{u} := y_t - \hat{B}^* x_t^* = y_t - \hat{B} x_t \quad,$$

where $\hat{B}$ is the unrestricted estimator of B and

$$\check{u} := y_t - \check{B}^* x_t^* = y_t - \check{B} x_t \quad,$$

where $\check{B}$ is the restricted estimator of B.

Therefore, the likelihood ratio λ is numerically identical if estimation is based on the transformed or untransformed model.

In order to define a test statistic with known exact distribution, the likelihood ratio will be transformed. Let $\mathcal{U}$ be the $(T/2)$th power of λ:

$$\mathcal{U} := \lambda^{2/T} = \frac{|\hat{\Sigma}|}{|\check{\Sigma}|} \quad . \tag{5.6}$$

Applying the following theorem given in Anderson (1984, p.305, theorem 8.4.5) allows the definition of a test statistic with known exact distribution:

The distribution of $[(1 - \mathcal{U}_{n,1,df})/\mathcal{U}_{n,1,df}] \cdot (df + 1 - n)/n$ follows an F-distribution with n and $df+1-n$ degrees of freedom. The subscript n denotes the number of equations, the subscript 1 denotes the number of restricted columns of B^* and df denotes the degrees of freedom with $df := T - (n + 3)$.

The statistic for testing homogeneity which is related to the $\mathcal{U}$ statistic (and by this to the likelihood ratio) will be called LRU and is defined as

$$LRU := \left(\frac{T - 2n - 2}{n} \right) \frac{1 - \mathcal{U}}{\mathcal{U}} \tag{5.7}$$

with

$$LRU \stackrel{H_0}{\sim} F_{n;(T-2n-2)} \quad .$$

The exact critical values of the LRU test are denoted by F_α and satisfy

$$P(F_{n;(T-2n-2)} \leq F_\alpha) = 1 - \alpha \quad . \tag{5.8}$$

The F-distribution is completely determined by its two degrees of freedom parameters. This implies that the distribution of the *LRU* statistic is independent of the x_t's, the coefficient matrix B^* (or B) and the covariance matrix Σ.

Invariance of the LRU test:

It should be noted that the LRU test is invariant to the equation deleted in the complete demand system. Recall that one equation of the complete demand system is deleted due to the adding-up restriction. The invariance property can be proven as follows: the likelihood values of the restricted (homogeneity) and unrestricted model are invariant to the equation deleted as shown in Barten (1969). The *LRU* statistic is a function of the likelihood values and is, therefore, invariant to the equation deleted.

5.2 Functional equivalence between the *LRU* and Laitinen's statistic

The distribution of the *LRU* statistic is identical to the distribution of the statistic proposed in Laitinen (1978) for testing homogeneity. One can show that the *LRU* statistic is just a reformulation of Laitinen's statistic. The most interesting point here is the functional equivalence between the *LRU* and Laitinen's statistic implying that these two statistics have the same distribution for any error process (if the distribution exists). Note that if two statistics have the same distribution, this does not necessarily imply that they are functionally equivalent. The motivation for demonstrating this functional equivalence is that this result can be used for studying the distribution of the *LRU* statistic under dynamic misspecification.

The procedure for proving the functional equivalence between the *LRU* and Laitinen's statistic is similar to Bera (1982), who showed the relation between the LR, Wald and LM test for testing homogeneity in an equation system with identical regressors and normal, homoskedastic errors which are independent in time.

Recall that the *LRU* statistic is defined as $[(T - 2n - 2)/n] \cdot (1 - \mathcal{U})/\mathcal{U}$ with $\mathcal{U} = |\hat{\Sigma}|/|\check{\Sigma}|$.

Proposition 5.1

The *LRU* statistic can be rewritten as

$$LRU = \left(\frac{T - 2n - 2}{(T - (n+3))n} \right) T(R_H \hat{b})' \left(R_H \hat{\Phi}^* R_H' \right)^{-1} (R_H \hat{b}) \quad ,$$

where

$$R_H := a' \otimes I_n \quad ,$$

with

$$a := \begin{pmatrix} 0_{2\times1} \\ \iota_{n+1} \end{pmatrix}$$

and

$$\hat{\varPhi}^* := M_T^{-1} \otimes \frac{\sum_{t=1}^T \hat{u}_t \hat{u}_t'}{T - (n+3)} \quad ,$$

with $M_T := \sum_{t=1}^T x_t x_t'/T$.

Proof:

Note that

$$\frac{1-\mathcal{U}}{\mathcal{U}} = \frac{1}{\mathcal{U}} - 1 = \frac{|\check{\Sigma}|}{|\hat{\Sigma}|} - 1 = \frac{|\sum_{t=1}^T \check{u}_t \check{u}_t'|}{|\sum_{t=1}^T \hat{u}_t \hat{u}_t'|} - 1 \quad ,$$

where $\check{u}_t$ denotes the restricted and $\hat{u}_t$ the unrestricted residuals. Next the term $\sum_{t=1}^T \check{u}_t \check{u}_t'$ will be expressed as a function of the unrestricted estimator $\hat{B}$.

The vector $\check{u}_t$ can be formulated as

$$\check{u}_t = y_t - \hat{B}x_t + \hat{B}x_t - \check{B}x_t = \hat{u}_t + (\hat{B} - \check{B})x_t$$

and by this

$$\sum_{t=1}^T \check{u}_t \check{u}_t' = \sum_{t=1}^T \hat{u}_t \hat{u}_t' + (\hat{B} - \check{B}) \sum_{t=1}^T x_t x_t' (\hat{B} - \check{B})' \quad , \qquad (5.9)$$

where the cross terms equal zero as $\hat{u}_t$ is orthogonal to x_t.

An expression of the restricted estimator $\check{B}$ as a function of the unrestricted estimator $\hat{B}$ can be found as follows: the restricted estimator of b is given by (see Bera 1982, p.292)

$$\check{b} = \hat{b} - \left((\sum_{t=1}^T x_t x_t')^{-1} \otimes \hat{\Sigma} \right) R_H' \left(R_H \left((\sum_{t=1}^T x_t x_t')^{-1} \otimes \hat{\Sigma} \right) R_H' \right)^{-1} R_H \hat{b} \quad ,$$

$$(5.10)$$

where $\hat{\Sigma} := \sum_{t=1}^T \hat{u}_t \hat{u}_t'/T$, $\hat{b} := \text{vec}\,\hat{B}$, $\check{b} := \text{vec}\,\check{B}$ and R_H is the restriction matrix such that $R_H \hat{b} = 0_{n\times1}$.

The restriction matrix R_H can be found as follows: recall that the homogeneity restriction was given by $(S\,|\,s)\iota_{n+1} = 0_{n\times1}$. Using the partition of $B = (C\,|\,S\,|\,s)$, and noting that homogeneity does not restrict the coefficient submatrix C, the homogeneity restriction imposed on the coefficient matrix B can be expressed as

$$(C\,|\,S\,|\,s) \begin{pmatrix} 0_{2\times1} \\ \iota_{n+1} \end{pmatrix} = 0_{n\times1} \quad .$$

This can be written more compactly as

$$Ba = 0_{n \times 1} \quad ,$$

where a is a $(n+3) \times 1$ vector defined by

$$a := \begin{pmatrix} 0_{2 \times 1} \\ \iota_{n+1} \end{pmatrix} \quad .$$

In vec notation, the homogeneity restriction takes the form

$$Ba = \text{vec } Ba = (a' \otimes I_n)\text{vec } B = 0_{n \times 1} \quad ,$$

where we have used the rule $\text{vec } IBC = (C' \otimes I)\text{vec } B$. Defining

$$R_H := (a' \otimes I_n) \quad , \tag{5.11}$$

which is a $n \times n(n+3)$ matrix, and using $b := \text{vec } B$, the homogeneity restriction in vec form can be written as

$$R_H b = 0_{n \times 1} \quad . \tag{5.12}$$

Replacing R_H in equation (5.10) by $(a' \otimes I_n)$ yields

$$\begin{aligned}
\check{b} &= \hat{b} - \left((\sum_{t=1}^{T} x_t x_t')^{-1} \otimes \hat{\Sigma} \right)(a \otimes I_n) \\
&\quad \left((a' \otimes I_n)\left((\sum_{t=1}^{T} x_t x_t')^{-1} \otimes \hat{\Sigma} \right)(a \otimes I_n) \right)^{-1}(a' \otimes I_n)\hat{b} \\
&= \hat{b} - \left((\sum_{t=1}^{T} x_t x_t')^{-1}a \otimes \hat{\Sigma} \right)\left((a'(\sum_{t=1}^{T} x_t x_t')^{-1}a)^{-1} \otimes \hat{\Sigma}^{-1} \right)(a' \otimes I_n)\hat{b} \\
&= \hat{b} - \left(\frac{(\sum_{t=1}^{T} x_t x_t')^{-1}aa' \otimes I_n}{a'(\sum_{t=1}^{T} x_t x_t')^{-1}a} \right)\hat{b} \quad . \tag{5.13}
\end{aligned}$$

Using this last result and the rule $(C' \otimes I_n)\text{vec } B = \text{vec } BC$, the restricted estimator $\check{B}$ can be expressed in matrix form as

$$\check{B} = \hat{B} - \frac{\hat{B}aa'(\sum_{t=1}^{T} x_t x_t')^{-1}}{a'(\sum_{t=1}^{T} x_t x_t')^{-1}a} \quad .$$

Now, the expression $\sum_{t=1}^{T} \check{u}_t \check{u}_t'$ in (5.9) can be reformulated as

$$\begin{aligned}
\sum_{t=1}^{T} \check{u}_t \check{u}_t' &= \sum_{t=1}^{T} \hat{u}_t \hat{u}_t' + (\hat{B} - \check{B})\sum_{t=1}^{T} x_t x_t'(\hat{B} - \check{B})' \\
&= \sum_{t=1}^{T} \hat{u}_t \hat{u}_t' + \frac{\hat{B}aa'(\sum_{t=1}^{T} x_t x_t')^{-1}(\sum_{t=1}^{T} x_t x_t')(\sum_{t=1}^{T} x_t x_t')^{-1}aa'\hat{B}'}{(a'(\sum_{t=1}^{T} x_t x_t')^{-1}a)^2} \\
&= \sum_{t=1}^{T} \hat{u}_t \hat{u}_t' + \frac{\hat{B}aa'\hat{B}'}{a'(\sum_{t=1}^{T} x_t x_t')^{-1}a} \quad . \tag{5.14}
\end{aligned}$$

Recalling that

$$\frac{1-\mathcal{U}}{\mathcal{U}} = \frac{1}{\mathcal{U}} - 1 = \frac{|\sum_{t=1}^{T} \breve{u}_t \breve{u}_t'|}{|\sum_{t=1}^{T} \hat{u}_t \hat{u}_t'|} - 1 \quad,$$

and replacing the expression for $\sum_{t=1}^{T} \breve{u}_t \breve{u}_t'$ as in (5.14) yields

$$\frac{1-\mathcal{U}}{\mathcal{U}} = \frac{\left| \sum_{t=1}^{T} \hat{u}_t \hat{u}_t' + \dfrac{\hat{B}aa'\hat{B}'}{a'(\sum_{t=1}^{T} x_t x_t')^{-1}a} \right|}{|\sum_{t=1}^{T} \hat{u}_t \hat{u}_t'|} - 1 \quad.$$

The numerator of the last equation can be simplified by applying the rule $|A + rr'| = |A|(1 + r'A^{-1}r)$, where A denotes a non-singular matrix and r a column vector (see Anderson 1984, Corollary A.3.1, p.594). Let $A = \sum_{t=1}^{T} \hat{u}_t \hat{u}_t'$ and $r = \hat{B}a/\sqrt{a'(\sum_{t=1}^{T} x_t x_t')^{-1}a}$. Then,

$$
\begin{aligned}
\frac{1-\mathcal{U}}{\mathcal{U}} &= \frac{|\sum_{t=1}^{T} \hat{u}_t \hat{u}_t'| \left(1 + \dfrac{a'\hat{B}'(\sum_{t=1}^{T} \hat{u}_t \hat{u}_t')^{-1}\hat{B}a}{a'(\sum_{t=1}^{T} x_t x_t')^{-1}a}\right)}{|\sum_{t=1}^{T} \hat{u}_t \hat{u}_t'|} - 1 \\[2ex]
&= \frac{a'\hat{B}'(\sum_{t=1}^{T} \hat{u}_t \hat{u}_t')^{-1}\hat{B}a}{a'(\sum_{t=1}^{T} x_t x_t')^{-1}a} \\[2ex]
&= \frac{\hat{b}'R_H'(\sum_{t=1}^{T} \hat{u}_t \hat{u}_t')^{-1}R_H\hat{b}}{a'(\sum_{t=1}^{T} x_t x_t')^{-1}a} \quad,
\end{aligned}
\tag{5.15}
$$

where, for the last equality, we have used $\hat{B}a = (a' \otimes I_n)\hat{b} = R_H\hat{b}$.

Now, using the definition of the LRU statistic and multiplying it by $\frac{T-(n+3)}{T-(n+3)}$ yields:

$$
\begin{aligned}
LRU &:= \left(\frac{T-2n-2}{n}\right)\left(\frac{1-\mathcal{U}}{\mathcal{U}}\right) \\[2ex]
&= \left(\frac{T-2n-2}{n}\right)\left(\frac{T-(n+3)}{T-(n+3)}\right)\left(\frac{\hat{b}'R_H'(\sum_{t=1}^{T} \hat{u}_t \hat{u}_t')^{-1}R_H\hat{b}}{a'(\sum_{t=1}^{T} x_t x_t')^{-1}a}\right) \\[2ex]
&= \left(\frac{T-2n-2}{(T-(n+3))n}\right)(\hat{b}'R_H')\left((a'(\sum_{t=1}^{T} x_t x_t')^{-1}a)\frac{\sum_{t=1}^{T} \hat{u}_t \hat{u}_t'}{T-(n+3)}\right)^{-1}(R_H\hat{b})
\end{aligned}
\tag{5.16}
$$

The reformulated LRU statistic in equation (5.16) is equivalent to the definition of Laitinen's statistic (1978, equation 5). (Note that Laitinen (1978) considers a model of $n-1$ equations and $n+1$ regressors, whereas we have n equations and $n+3$ regressors.)

The last equation will be written more compactly for the analysis below. The third term on the right-hand side of equation (5.16) can be written as

$$\left(a'(\sum_{t=1}^{T} x_t x_t')^{-1} a\right)^{-1} \otimes \left(\frac{\sum_{t=1}^{T} \hat{u}_t \hat{u}_t'}{T - (n+3)}\right)^{-1} =,$$

$$(a' \otimes I_n)\left((\sum_{t=1}^{T} x_t x_t')^{-1} \otimes \frac{\sum_{t=1}^{T} \hat{u}_t \hat{u}_t'}{T - (n+3)}\right)^{-1} (a \otimes I_n)$$

Define

$$\hat{\Phi}^* := M_T^{-1} \otimes \frac{\sum_{t=1}^{T} \hat{u}_t \hat{u}_t'}{T - (n+3)} \quad ,$$

where $M_T := \sum_{t=1}^{T} x_t x_t'/T$. Using $R_H := a' \otimes I_n$, the *LRU* statistic can be formulated as

$$LRU = \left(\frac{T - 2n - 2}{(T - (n+3))n}\right) T(R_H \hat{b})' \left(R_H \hat{\Phi}^* R_H'\right)^{-1} (R_H \hat{b}) \quad , \qquad (5.17)$$

which is the intended result.

5.3 Likelihood ratio test if the errors are VAR(p)

In this section, the likelihood ratio statistic for testing homogeneity with normal, homoskedastic but autocorrelated errors is considered. The exact distribution of the test statistic is unknown, and the small sample critical value of the likelihood ratio test can differ substantially from its asymptotical critical value. In order to take this small sample problem into account, two different testing procedures are discussed. First, a modified likelihood ratio statistic with a small sample correction factor will be proposed. Second, a Monte Carlo test will be defined.

5.3.1 The likelihood ratio test (LR)

The transformed model equation is defined by $y_t = B^* x_t^* + u_t$ with the following assumption:

Assumption 5.2

a) $u_t = \sum_{j=1}^{p} R_j u_{t-j} + \epsilon_t$, where u_t is a $n \times 1$ vector of autocorrelated errors, the u_{t-j} are $n \times 1$ vectors of lagged errors and the R_j are a $n \times n$ matrices of autoregression coefficients.

b) ϵ_t is a $n \times 1$ vector which is distributed as

$$\epsilon_t \sim i.i.d.N(O, \Sigma) \qquad \forall t \quad .$$

The hypotheses for testing homogeneity are

$$H_0: \quad s^* = 0_{n \times 1} \quad \text{and} \quad H_1: \quad s^* \neq 0_{n \times 1} \quad .$$

The likelihood ratio $\overline{\lambda}$ is defined by

$$\overline{\lambda} := \frac{\displaystyle\max_{B^*, \Sigma, R \in H_0} L_{\text{VAR}}(B^*, \Sigma, R)}{\displaystyle\max_{B^*, \Sigma, R \in H_0 \cup H_1} L_{\text{VAR}}(B^*, \Sigma, R)} \quad , \tag{5.18}$$

where $L_{\text{VAR}}(\cdot)$ is the likelihood function related to the previous model and where $\mathcal{R} := (R_1, R_2, \dots, R_p)$.

The log-likelihood function for a multivariate regression model with normal homoskedastic errors following a VAR(p) process can be found as follows: Recall that $u_t = \sum_{j=1}^{p} R_j u_{t-j} + \epsilon_t$ where ϵ_t is normal, homoskedastic and independent in time. Using the relations $\epsilon_t = u_t - \sum_{j=1}^{p} R_j u_{t-j}$ and $u_{t-j} = y_{t-j} - B^* x_{t-j}^*$, we can express ϵ_t as

$$\begin{aligned}
\epsilon_t &= u_t - \sum_{j=1}^{p} R_j u_{t-j} = (y_t - B^* x_t^*) - \sum_{j=1}^{p} R_j (y_{t-j} - B^* x_{t-j}^*) \\
&= y_t - B^* x_t^* - \sum_{j=1}^{p} R_j y_{t-j} + \sum_{j=1}^{p} R_j B^* x_{t-j}^* \quad .
\end{aligned} \tag{5.19}$$

Using the last equation, and recalling that ϵ_t follows a multivariate normal distribution, the log-likelihood function conditional on the first p observations can be written as

$$\ln L_{\text{VAR}}(B^*, \mathcal{R}, \Sigma | y_0, y_{-1}, \dots y_{-p+1}) = -\frac{nT}{2} \ln 2\pi - \frac{T}{2} \ln |\Sigma| - \sum_{t=1}^{T} \frac{1}{2} \text{tr} \, \Sigma^{-1} \epsilon_t \epsilon_t'$$

with ϵ_t being defined in (5.19).

The formulas for calculating the conditional maximum likelihood estimators with and without homogeneity imposed are given in Deschamps (1993, Theorem 1). As in the previous section, the log-likelihood function has been formulated for the transformed model in order to emphasise that homogeneity restricts one column of the coefficient matrix B^*. As the underlying probability model is the same for the transformed and untransformed model, the numerical value of $\overline{\lambda}$ is the same for both types of models. Under the null hypothesis, the asymptotic distribution of the usual monotonic transformation of $\overline{\lambda}$ denoted by LR is given by

$$LR := -2 \ln \overline{\lambda} \overset{d}{\to} \chi_n^2 \quad , \tag{5.20}$$

where the subscript n denotes the degrees of freedom. The critical values of the LR test are denoted by χ_α and satisfy

$$P(\chi_n^2 \leq \chi_\alpha) = 1 - \alpha \quad .$$

The small sample distribution of this likelihood ratio statistic is not known, but is well known that the asymptotic distribution of LR may be a poor approximation of its small sample distribution. This implies that for a given size of the test, the small sample critical values will deviate from the asymptotic critical values, where the latter are simply based on the χ^2 distribution. The homogeneity test can, therefore, be biased substantially in small samples.

5.3.2 Small sample corrected likelihood ratio test (LRC)

We propose to use a modified likelihood ratio statistic which is related to Anderson's $\mathcal{U}$ statistic.

The modified statistic is denoted by LRC and defined as

$$LRC := \left(\frac{T - 2n - 2}{n} \right) \frac{1 - \overline{\mathcal{U}}}{\overline{\mathcal{U}}} \quad , \tag{5.21}$$

where

$$\overline{\mathcal{U}} := \overline{\lambda}^{2/T} \quad .$$

The critical values of the LRC test are denoted by F_α and satisfy

$$P(F_{n;(T-2n-2)} \leq F_\alpha) = 1 - \alpha \quad .$$

The motivation for the LRC statistic is as follows: in section 5.1, Anderson's $\mathcal{U}$ test was applied to a linear multivariate regression model with normal, homoskedastic errors which are independent in time and where one column of the coefficient matrix is restricted. Recall that homogeneity restricts one column of the coefficient matrix B^* as shown above. Here the errors follow a VAR(p) process and, therefore, the errors are not independent. Heuristically, we distinguish two sorts of biases. One bias is due to small sample estimation independent of the values of the autoregression coefficients $\mathcal{R}$. The use of the LRC statistic here should be seen as a correction for this source of bias. The other bias is related to the fact that $R \neq (O, O, \ldots, O)$ which may increase or decrease the size effect related to small sample estimation. Clearly these two biases may not be independent, but this approach helps to distinguish between the two sorts of biases in Monte Carlo experimentation.

The relation between the LRC and LR statistic is characterised by the following proposition:

Proposition 5.2

a) The LRC statistic is functionally related to the LR statistic by

$$LRC = \left(\frac{T - 2n - 2}{n}\right)(\exp(LR/T) - 1)$$

b) The size of the LRC test is smaller than the size of the LR test if

$$T \ln\left(\frac{F_\alpha n}{T - 2n - 2} + 1\right) > \chi_\alpha \quad,$$

where F_α denotes the critical value of the LRC test and χ_α the critical values of the LR test.

c) The statistics LRC and LR are asymptotically equivalent. Proposition 5.2 b) gives the condition for which the the null hypothesis of homogeneity is rejected less often with the LRC test than with the LR test in finite sample for given α.

Proof

For proving proposition 5.2 a), the LRC statistic will be formulated as a function of $\bar\lambda$.
Using $\overline{U} := \bar\lambda^{2/T}$, it follows that

$$LRC := \left(\frac{T - 2n - 2}{n}\right)\frac{1 - \overline{U}}{\overline{U}} = \left(\frac{T - 2n - 2}{n}\right)(\bar\lambda^{-2/T} - 1) \quad.$$

Next, the likelihood ratio $\bar\lambda$ will be expressed as a function of the LR statistic. Recall that $LR := -2\ln\bar\lambda$. It follows directly that

$$\bar\lambda = \exp(-LR/2) \quad.$$

Combining the last two equations yields

$$LRC = \left(\frac{T - 2n - 2}{n}\right)(\exp(LR/T) - 1) \quad,$$

which proves proposition 5.2 a).

For proving proposition 5.2 b), we derive the conditions for which

$$Pr(LRC \leq F_\alpha) > Pr(LR \leq \chi_\alpha)$$

holds.

Using proposition 5.2 a), one can rewrite the left-hand side of the inequality above as

$$
\begin{aligned}
Pr(LRC \leq F_\alpha) &= Pr\left((\frac{T - 2n - 2}{n})(\exp(LR/T) - 1) \leq F_\alpha\right) \\
&= Pr\left(LR \leq T\ln(\frac{F_\alpha n}{T - 2n - 2} + 1)\right) \quad. \qquad (5.22)
\end{aligned}
$$

The size of the LRC test is smaller than the size of the LR test if

$$Pr(LRC \leq F_\alpha) > Pr(LR \leq \chi_\alpha)$$

or equivalently from equation (5.22) if

$$Pr\left(LR \leq T\ln(\frac{F_\alpha n}{T - 2n - 2} + 1)\right) > Pr\left(LR \leq \chi_\alpha\right) \quad .$$

The last inequality is satisfied if

$$T\ln\left(\frac{F_\alpha n}{T - 2n - 2} + 1\right) > \chi_\alpha \quad .$$

This proves proposition 5.2 b).

The proof for proposition 5.2 c) is omitted (see Anderson 1984, ch. 8.5).

5.3.3 A Monte Carlo test (LR-MC)

In this section, a Monte Carlo test for homogeneity is proposed. This is an alternative to the likelihood ratio test discussed above. The motivation for this is that the distribution of the likelihood ratio statistic (LR) discussed above is only known asymptotically. Statistical inference from the LR test is, therefore, only justified asymptotically and often inadequate in small samples. A first possible remedy was proposed in the last section. Monte Carlo testing is a different approach to deal with the small sample problem. Today, this kind of simulation-based inference is feasible in many applications since the computational costs have fallen dramatically. The idea of Monte Carlo testing goes back to Dwass (1957) and Barnard (1961). Monte Carlo testing can be thought of as a parametric bootstrap test. The procedure of Monte Carlo testing consists in estimating the distribution of the test statistic under the null hypothesis by simulation for an assumed data-generating process. The assumed data-generating process is based on the restricted estimates of the sample model. The original Monte Carlo test proposed in Dwass (1957) and Barnard (1961) does *not* depend on unknown parameters. An example is the traditional *t*-statistic for the univariate regression model under the appropriate assumptions. Then the Monte Carlo test generates simply realisations of the student distribution. The Monte Carlo test considered here does depend on unknown (nuisance) parameters. There already exist many versions of bootstrap tests in literature. Here only one specific version of the bootstrap test is considered; this is Barnard's type of bootstrap testing which will be applied to the LR test. It should be noted that this type of bootstrap testing is different from the parametric bootstrap test proposed in Hall and Titterington (1989) for example. A brief overview of some popular versions of bootstrap tests in applied econometrics is given in Veall (1998) and Horowitz (1996), e.g. A rigourous justification for the bootstrap approach is given in Hall (1992). The theoretical argument for why the bootstrap works is based on asymptotic expansions of the test statistic which provide an asymptotic refinement. This is a very technical and tedious matter. It should be noted that in many situations an asymptotic expansion of the test statistic does not exist. Nevertheless, bootstrap testing can perform better than traditional testing procedures in these cases. Furthermore, even if the theoretical rigourous justification of the bootstrap test holds, this does not imply that it works well in empirical work. The performance of bootstrap tests should, therefore, always be analysed by case studies.

Here the interesting case is the omitted variable test for the multivariate regression model with normally distributed errors following a VAR(p) process. For multivariate regression systems, Monte Carlo tests have been applied by Taylor, Shonkwiler and Theil (1986) for example. They apply Barnard's type of bootstrap test for demand homogeneity. They assume that the errors are normal, homoskedastic and independent in time and use a non-standard test statistic. For the same kind of model, Cribari-Neto and Zarkos (1997) analyse the bootstrap test for demand homogeneity related to the Wald, LM and likelihood ratio test. For univariate regression models with AR(1) errors, bootstrap confidence intervals have been analysed in Rilstone (1993) for example. For the kind of model considered here, the performance of the Monte Carlo test has not yet been analysed to our knowledge.

The model is given by $y_t = B x_t + u_t$ with $u_t = \sum_{j=1}^{p} R_j u_{t-j} + \epsilon_t$, where $\epsilon_t \sim i.i.d.N(O, \Sigma)$. The hypotheses are given by $H_0 : Ba = O$ and $H_1 : Ba \neq O$. The Monte Carlo test is implemented as follows:

1. Estimate the model under the null hypothesis of homogeneity. The lag order p of the error process is assumed to be known.
2. Calculate the LR statistic: LR.
3. Use the constrained estimates for $(B, \Sigma, R_1, R_2, \ldots, R_p)$ for generating (bootstrapping) the artificial data set j: (y_t^{*j}, u_t^{*j}).
4. Calculate the likelihood ratio statistic based on the artificial data set j: LR^{*j}
5. Repeat step 3-4 N_B times.
6. Estimate the critical value from the sample distribution of LR^{*j} for given nominal size α: LR_α^*.
7. If $LR > LR_\alpha^*$, reject the null hypothesis, otherwise not.

It should be noted that the distribution of the LR statistic depends on the parameters $(B, \Sigma, R_1, R_2, \ldots, R_p)$. Since the true parameter values are not known and replaced by their estimates under the null, the Monte Carlo test is not exact. This means that the true type I error is not equal to the nominal type I error in general. The performance of the Monte Carlo test will, therefore, be investigated in the simulation study.

5.4 The distribution of the *LRU* statistic under dynamic misspecification

In this section the distribution of the *LRU* statistic under dynamic misspecification is studied. In sections 5.1 and 5.3, tests for homogeneity were defined under the assumption that the error process is correctly specified. The exact distribution of the *LRU* statistic under the null hypothesis of homogeneity is known under the assumptions of normal homoskedastic errors which are independent in time. An interesting question concerns the consequences of using the LRU test if this assumption does not hold; for example if the errors are *not* independently distributed. The analysis is applied to a restricted form of dynamic misspecification which facilitates the interpretation of the results. For this restricted class of models, the asymptotic distribution of the Wald statistic is derived. The empirical motivation is the homogeneity test for a consumer demand system under dynamic misspecification.

For univariate regression models, the distribution of some statistics have been analysed where the error process is dynamically misspecified. For special cases, the exact distribution of the misspecified statistic was found. For example, Giles and Scott (1992) found the exact distribution of the traditional F-statistic for the Chow test if the errors follow a normal stationary AR(1) process; the exact distribution is a quadratic form in normal variables.

For the multivariate regression model, however, the omitted variable test has not yet been analysed under dynamic specification to our knowledge. Here we are interested in the distribution of the *LRU* statistic under dynamic misspecification. It will be shown that the *LRU* statistic is distributed asymptotically as a quadratic form in normal variables under quite general assumptions. For the multivariate regression model, small sample results of the kind given in Giles and Scott (1992) seem not to be analytically tractable.

Recall that the *LRU* statistic could be reformulated as (see equation (5.17))

$$LRU = \left(\frac{T - 2n - 2}{(T - (n+3))n} \right) T\hat{b}' R'_H \left(R_H \hat{\Phi}^* R'_H \right)^{-1} R_H \hat{b} \quad ,$$

where the $n \times n(n+3)$ restriction matrix R_H was defined as

$$R_H := a' \otimes I_n \quad ,$$

the $n(n+3) \times 1$ parameter vector b is estimated by

$$\hat{b} = \text{vec} \sum_{t=1}^{T} y_t x'_t \left(\sum_{t=1}^{T} x_t x'_t \right)^{-1} \quad .$$

and the $n(n+3) \times n(n+3)$ variance-covariance matrix of $T^{1/2}(\hat{b} - b)$ is estimated by

$$\hat{\Phi}^* := \left(M_T^{-1} \otimes \sum_{t=1}^{T} \frac{\hat{u}_t \hat{u}_t'}{T - (n+3)} \right)$$

with

$$M_T := \sum_{t=1}^{T} \frac{x_t x_t'}{T} \quad .$$

Note that the estimator $\hat{b}$ can be interpreted as a GMM estimator. In section 4.1, it was mentioned that the GMM estimator is consistent under quite general assumptions. Here the problem of wrongly specifying the error process is related to estimating the variance-covariance matrix of $T^{1/2}(\hat{b} - b)$ by an estimator which is only consistent in special cases.

In order to derive the asymptotic distribution of the LRU statistic if the error process is wrongly specified, the following assumption is made:

Assumption 5.3

a) $T^{1/2}(\hat{b} - b) \overset{d}{\to} N(0, \Phi)$, where b denotes the true parameter vector and the asymptotic variance of $T^{1/2}(\hat{b} - b)$ is given by

$$\Phi := (M^{-1} \otimes I_n) V (M^{-1} \otimes I_n)$$

where

$$M_T := \sum_{t=1}^{T} \frac{x_t x_t'}{T} \overset{p}{\to} M$$

and

$$V := \lim_{T \to \infty} Var(T^{-1/2} \sum_{t=1}^{T} v_t) \quad \text{with} \quad v_t := x_t \otimes u_t$$

b) Φ is estimated by

$$\hat{\Phi}^* := M_T^{-1} \otimes \sum_{t=1}^{T} \frac{\hat{u}_t \hat{u}_t'}{T - (n+3)} \quad ,$$

which is an inconsistent estimator in general.

c) $\sum_{t=1}^{T} u_t u_t' / T \overset{p}{\to} \Sigma_u$ with Σ_u being finite and positive definite.

We will show that the LRU statistic is asymptotically equivalent to a quadratic form in normal variables and its distribution can be calculated numerically by using a result from Imhof (1961).

Proposition 5.3

Under the null hypothesis of homogeneity, the asymptotic distribution of the *LRU* statistic can be characterised by

$$P(LRU \leq e) \overset{a}{=} P(\frac{1}{n} \sum_{i=1}^{n} \lambda_i z_i^2 \leq e) \quad ,$$

where $\overset{a}{=}$ means asymptotically equivalent, z_i is $i.i.d. N(0,1)$ for $i = 1, 2, \ldots, n$ and

$$\lambda_i = e_i \left((a'M^{-1}a)^{-1} \Sigma_u^{-1} (R_H \Phi R_H') \right) \quad ,$$

where $e_i(\cdot)$ denotes the ith eigenvalue of the term in brackets.

Proof

In order to find the asymptotic distribution of *LRU*, the convergence results for the different terms of the reformulated *LRU* statistic are derived. Recall that the *LRU* statistic could be written as

$$\left(\frac{T - 2n - 2}{(T - n - 3)n} \right) \left(T^{1/2} R_H \hat{b} \right)' \left(R_H \hat{\Phi}^* R_H' \right)^{-1} \left(T^{1/2} R_H \hat{b} \right) \quad . \tag{5.23}$$

For the first term of equation (5.23), note that

$$\frac{T - 2n - 2}{(T - n - 3)n} \to \frac{1}{n} \qquad \text{for } T \to \infty \quad . \tag{5.24}$$

The second term and last term converge in distribution.

By assumption 5.3 a) $T^{1/2}(\hat{b} - b) \overset{d}{\to} N(O, \Phi)$ and under the null hypothesis of homogeneity $R_H b = O$. Therefore,

$$T^{1/2} R_H \hat{b} \overset{d}{\to} N(O, R_H \Phi R_H') \quad .$$

Finally, the third term converges in probability. In order to show this, the probability limit of

$$\hat{\Phi}^* := M_T^{-1} \otimes \sum_{t=1}^{T} \frac{\hat{u}_t \hat{u}_t'}{(T - n - 3)}$$

will be derived.

By assumption $M_T \overset{p}{\to} M$ and, therefore, $M_T^{-1} \overset{p}{\to} M^{-1}$.

The term $\sum_{t=1}^{T} \hat{u}_t \hat{u}_t' / (T - n - 3)$ has the same probability limit as $\sum_{t=1}^{T} \hat{u}_t \hat{u}_t' / T$. Using exactly the same proof as in section 4.1, one can show that

$$\sum_{t=1}^{T} \frac{\hat{u}_t \hat{u}_t'}{T} - \sum_{t=1}^{T} \frac{u_t u_t'}{T} \overset{p}{\to} O \quad ,$$

which implies that

$$\sum_{t=1}^{T} \frac{\hat{u}_t \hat{u}_t'}{T} \overset{p}{\to} \Sigma_u \quad .$$

Using these results, it follows that

$$\hat{\Phi}^* \overset{p}{\to} M^{-1} \otimes \Sigma_u \quad .$$

Then,

$$R_H \hat{\Phi}^* R_H' \overset{p}{\to} (a' \otimes I_n)(M^{-1} \otimes \Sigma_u)(a \otimes I_n) = a' M^{-1} a \otimes \Sigma_u = a' M^{-1} a \Sigma_u, \tag{5.25}$$

where the last step follows from the fact that $a' M^{-1} a$ is a scalar.

It was shown above that the first term of the LRU statistic in equation (5.23) converges to a constant scalar, the second and fourth term are asymptotically normally distributed and the third term converges in probability to a constant matrix. Making use of these results, it can be seen that the LRU statistic in equation (5.23) is asymptotically equivalent to a quadratic form in normal variables. Next, it will be reformulated as a quadratic form in normal variables with unit variance.

Note that $R_H \Phi R_H'$ is a symmetric matrix of rank n, which can be diagonalised by the spectral decomposition theorem:

$$R_H \Phi R_H' = C \Delta C' \quad . \tag{5.26}$$

Recalling that $T^{1/2} R_H \hat{b} \overset{d}{\to} N(O, R_H \Phi R_H')$, it follows that

$$T^{1/2} \Delta^{-1/2} C' R_H \hat{b} \overset{d}{\to} N(O, I_n) \quad .$$

Premultiplying $\hat{b}$ in equation (5.23) by $(C \Delta^{1/2})(\Delta^{-1/2} C')$ (which equals I_n) and using $(T - 2n - 2)/(T - n - 3)n \to 1/n$ for $T \to \infty$ and $R_H \hat{\Phi}^* R_H' \overset{p}{\to} a' M^{-1} a \Sigma_u$, the LRU statistic is asymptotically equivalent to

$$LRU \overset{a}{=} \frac{1}{n} \left(T^{1/2} \hat{b}' R_H' C \Delta^{-1/2} \right) \left(\Delta^{1/2} C' (a' M^{-1} a \Sigma_u)^{-1} C \Delta^{1/2} \right) \\ \left(T^{1/2} \Delta^{-1/2} C' R_H \hat{b} \right) \quad , \tag{5.27}$$

where $\overset{a}{=}$ means asymptotically equivalent.

Note that the middle term of the last equation is also symmetric and of rank n, so that it can be diagonalised by the spectral decomposition theorem:

$$\Delta^{1/2} C' (a' M^{-1} a \Sigma_u)^{-1} C \Delta^{1/2} = P \Lambda P' \quad . \tag{5.28}$$

Let z denote a $n \times 1$ random vector distributed as $N(O, I_n)$, which is asymptotically equivalent to $T^{1/2} \Delta^{-1/2} C' R_H \hat{b}$. Furthermore, note that $P'z$ is also distributed as $N(O, I_n)$, which follows from the orthogonality of the matrix P, implying that $P'P = I_n$.

Using the definition of z and equations (5.27) and (5.28), one can conclude that

$$LRU \overset{a}{=} \frac{1}{n} z' \Lambda z = \frac{1}{n} \sum_{i=1}^{n} \lambda_i z_i^2 \quad ,$$

where λ_i is the ith diagonal element of Λ and z_i is the ith element of the vector z. The random variable z_i^2 is distributed as a central χ^2 with one degree of freedom.

The distribution of quadratic form in normal variables is given in Imhof (1961, p.422, equation 3.2). Using Imhof's formula, one can conclude that

$$\Pr(LRU \leq e) \approx \Pr(\frac{1}{n} \sum_{i=1}^{n} \lambda_i z_i^2 \leq e) = \Pr(\sum_{i=1}^{n} \lambda_i z_i^2 \leq e \cdot n)$$

$$= \frac{1}{2} - \frac{1}{\pi} \int_{0}^{\infty} \frac{\sin \theta(u)}{u \rho(u)} du \quad , \tag{5.29}$$

where e is a constant,

$$\theta(u) := \frac{1}{2} \sum_{i=1}^{n} \left(\tan^{-1}(\lambda_i u) \right) - \frac{1}{2}(e \cdot n)u$$

and

$$\rho(u) := \prod_{i=1}^{n} (1 + \lambda_i^2 u^2)^{1/4} \quad .$$

The definitions of $\theta(u)$ and $\rho(u)$ are slightly simplified compared to formulas given in Imhof (1961) as the z_i^2's here are independent central χ^2 variables with one degree of freedom. For given numerical values of the λ_i's, the integral in equation (5.29) can be calculated by using the algorithm given in Davies (1984).

In order to interpret the results, we will show that all the eigenvalues λ_i are positive, depending on the matrices Σ_u, M and V and taking the value one if the asymptotic variance Φ is estimated consistently.

These properties can be proven as follows:
Let

$$F := \Delta^{1/2} C' \left((a' M^{-1} a)^{-1} \Sigma_u^{-1} \right) C \Delta^{1/2}$$

and $e_i(F)$ be the ith eigenvalue of the matrix F. Note that the λ_i's are just the eigenvalues of F (see equation (5.28)).

Property 1: $\lambda_i > 0$ for $i = 1, \ldots n$.

It will be shown that the matrix F is positive definite which implies that the λ_i's are all strictly positive. We will make use of the rule that if A is a $n \times n$ positive definite matrix and B is any $n \times m$ matrix, then $B'AB$ is non-negative definite.

By assumption 5.3 c), the matrix Σ_u is positive definite implying that its inverse is also positive definite. The matrix M is also positive definite by assumption and, therefore, its inverse. It follows directly that the quadratic form $a'M^{-1}a$ is a positive scalar. These results imply that the middle term of F, which is $(a'M^{-1}a)^{-1}\Sigma_u^{-1}$, is positive definite. Premultiplying the matrix $(a'M^{-1}a)^{-1}\Sigma_u^{-1}$ by $\Delta^{1/2}C'$ and postmultiplying it by its transpose yields a non-negative definite matrix by the rule mentioned above. As F has full rank, one can conclude that F is positive definite and, therefore, all λ_i's are strictly positive. This property implies that $z'\Lambda z/n$ is a positive definite quadratic form.

Property 2: λ_i generally depends on Σ_u, M and V.

Using the rule $e_i(F) = e_i(A^{-1}FA)$ and recalling that the spectral decomposition of $R_H\Phi R_H'$ was given as $R_H\Phi R_H' = C\Delta C'$ (see equation (5.26)), it follows that

$$
\begin{aligned}
e_i(F) &= e_i\left(\Delta^{1/2}C'(a'M^{-1}a)^{-1}\Sigma_u^{-1}C\Delta^{1/2}\right) \\
&= e_i\left((a'M^{-1}a)^{-1}\Sigma_u^{-1}(C\Delta C')\right) \\
&= e_i\left((a'M^{-1}a)^{-1}\Sigma_u^{-1}(R_H\Phi R_H')\right) \quad .
\end{aligned}
\tag{5.30}
$$

Recall that $R_H := a' \otimes I_n$ and that the asymptotic variance Φ was given by

$$
(M^{-1} \otimes I_n)V(M^{-1} \otimes I_n) \quad .
$$

Using these definitions, $e_i(F)$ can be expressed as

$$
\begin{aligned}
e_i(F) &= e_i\left(((a'M^{-1}a)^{-1}\Sigma_u^{-1})(a' \otimes I_n)(M^{-1} \otimes I_n)V(M^{-1} \otimes I_n)(a \otimes I_n)\right) \\
&= e_i\left(((a'M^{-1}a)^{-1} \otimes \Sigma_u^{-1})(a'M^{-1} \otimes I_n)V(M^{-1}a \otimes I_n)\right) \\
&= e_i\left((a'M^{-1}a)^{-1}a'M^{-1} \otimes \Sigma_u^{-1}\right)V\left(M^{-1}a \otimes I_n\right) \quad ,
\end{aligned}
\tag{5.31}
$$

which cannot be simplified without further assumptions on the form of V.

Property 3:

If Φ is estimated consistently, λ_i equals one for all i.
In section 4.1, it was shown that the estimator

$$
\hat{\Phi} = M_T^{-1} \otimes \sum_{t=1}^{T} \frac{\hat{u}_t\hat{u}_t'}{T}
$$

is consistent if the errors are homoskedastic and serially uncorrelated. Recalling that $\hat{\Phi}^*$ equals $\hat{\Phi}$ if T is replaced by $T - (n+3)$ in the equation above, this implies that $\hat{\Phi}^*$ is also consistent for this special case.
Consistent estimation of Φ implies that

$$R_H \hat{\Phi}^* R'_H \xrightarrow{P} R_H \Phi R'_H \quad . \tag{5.32}$$

Furthermore, it was shown in equation (5.25) that generally

$$R_H \hat{\Phi}^* R_H \xrightarrow{P} a' M^{-1} a \Sigma_u \quad .$$

Therefore, if $\hat{\Phi}^*$ is a consistent estimator,

$$R_H \Phi R'_H = a' M^{-1} a \Sigma_u \quad . \tag{5.33}$$

Using the last equality, one can show that the eigenvalues of F all equal one. Recall that by equation (5.30),

$$e_i(F) = e_i \left((a' M^{-1} a)^{-1} \Sigma_u^{-1} R_H \Phi R'_H \right) \quad .$$

From equation (5.33), it follows directly that

$$e_i(F) = e_i(I_n) = 1 \quad ,$$

which holds for $i = 1, 2 \ldots n$.

Therefore, if Φ is estimated consistently, the λ_i's equal one and under the null hypothesis of homogeneity

$$LRU \stackrel{a}{=} \frac{1}{n} \sum_{i=1}^{n} z_i^2 \quad ,$$

implying that LRU is asymptotically distributed as a χ_n^2 divided by n (or equivalently as a central F-distribution with n and ∞ degrees of freedom parameters). If Φ is an inconsistent estimator, the asymptotic distribution of LRU depends on M, V and Σ_u through the definition of the λ_i's.

Approximation of the distribution of the *LRU* statistic by numerical integration

The nominal critical value of the LRU test is given by F_α satisfying $P(F_{n;T-2n-2} \leq F_\alpha) = 1 - \alpha$. The asymptotic distribution of the LRU statistic can be used to approximate the small sample distribution of the LRU statistic. From above we know that

$$\Pr(LRU \leq F_\alpha) \approx \Pr(\frac{1}{n} \sum_{i=1}^{n} \lambda_i z_i^2 \leq F_\alpha) \quad . \tag{5.34}$$

Here we propose a slightly different approximation of the small sample distribution of the LRU statistic. This approximation consists in replacing F_α on the right-hand side of (5.34) by χ_α/n, where $P(\chi_n^2 \leq \chi_\alpha) = 1 - \alpha$:

$$\Pr(LRU \leq F_\alpha) \approx \Pr(\frac{1}{n} \sum_{i=1}^{n} \lambda_i z_i^2 \leq \frac{\chi_\alpha}{n}) = \Pr(\sum_{i=1}^{n} \lambda_i z_i^2 \leq \chi_\alpha) \quad . \tag{5.35}$$

Note that the two approximations in (5.34) and (5.35) are is asymptotically equivalent since $\lim_{T\to\infty} F'_\alpha = \chi_\alpha/n$. The advantage of using the second approximation formula in (5.35) is that it is exact if the LRU test were correctly specified. This follows from property 3 on page 66 which states the consistent estimation of the variance-covariance matrix implies that $\lambda_i = 1$ for $i = 1, 2, \ldots, n$. In this case,

$$\Pr(LRU \le F_\alpha) = \Pr(\sum_{i=1}^{n} z_i^2 \le \chi_\alpha) = 1 - \alpha \quad .$$

Examples

We will now consider two illustrative examples which will be studied further in chapter 6. For both examples the regressors x_t are assumed to be non-stochastic. In the first example, it will be assumed that the errors u_t follow a covariance-stationary VAR(1) and in the second example a vector MA(1) process[1]. These examples are of interest because they are related to the data-generating processes used in the simulation study and will be analysed as follows: first, a set of assumptions is given in order to simplify the analysis. From these assumptions a simplified expression for calculating $\lambda_i = e_i(F)$ is derived and applied to our examples.

Assumption 5.4

a) u_t is covariance-stationary with absolutely summable autocovariances. The autocovariances are defined by

$$\Sigma_{u,j} := E(u_t u'_{t-j}) \quad \forall j, t \quad .$$

b) The matrices

$$M_j := \lim_{T\to\infty} E(\sum_{t=j+1}^{T} \frac{x_t x'_{t-j}}{T}) \quad \forall j$$

are positive definite and finite.

c) $E(u_t \otimes x_\tau) = 0_{n(n+3)\times 1}$ for any t and τ.

d) The asymptotic variance of $T^{-1/2} \sum_{t=1}^{T} v_t$ with $v_t := x_t \otimes u_t$ is denoted by V and is given by

$$V := \lim_{T\to\infty} Var(T^{-1/2} \sum_{t=1}^{T} v_t) \quad .$$

---

[1] A MA(1) process denotes a moving average process of order 1.

Proposition 5.4

If assumption 5.4 holds, the eigenvalues λ_i which are generally given by

$$\lambda_i := e_i(F) = e_i \left((a'M_0^{-1}a)^{-1} \Sigma_u^{-1} R_H \Phi R_H' \right)$$

can be calculated as

$$e_i \left(I_n + \sum_{j=1}^{\infty} \frac{a'M_0^{-1}M_j M_0^{-1}a}{a'M_0^{-1}a} \Sigma_{u,0}^{-1}(\Sigma_{u,j} + \Sigma_{u,j}') \right) \qquad . \tag{5.36}$$

Proof
See Appendix 1.

For the following two examples, it is assumed that assumption 5.4 holds in general so that proposition 5.4 can be applied.

Example 5.1

In this example, the regressors x_t are non-stochastic and the errors u_t follow a covariance-stationary VAR(1) process defined by

$$u_t = Ru_{t-1} + \epsilon_t$$

with

$$E(\epsilon_t) = 0$$

and

$$E(\epsilon_t \epsilon_{t-j}') = \begin{cases} \Sigma & \text{if } j = 0 \\ O & \text{otherwise} \end{cases} \qquad .$$

For applying proposition 5.4, one needs the expression for $\Sigma_{u,j}$. Recall that $\Sigma_{u,j} := E(u_t u_{t-j}')$. First, $\Sigma_{u,0}$ will be calculated as outlined in Hamilton (1994, p.266). Then, $\Sigma_{u,j}$ will be expressed as a function of R and $\Sigma_{u,0}$. Using the equation $u_t = Ru_{t-1} + \epsilon_t$, postmultiplying the left and right side by its transpose and taking expectations yields:

$$\begin{aligned} \Sigma_{u,0} := E(u_t u_t') &= E\left(Ru_{t-1} + \epsilon_t \right)\left(Ru_{t-1} + \epsilon_t \right)' \\ &= RE(u_{t-1}u_{t-1}')R' + RE(u_{t-1}\epsilon_t') + RE(\epsilon_t u_{t-1}') + E(\epsilon_t \epsilon_t'). \end{aligned} \tag{5.37}$$

Covariance-stationarity of u_t implies that its second moments are time-invariant and, therefore, $E(u_t u_t') = E(u_{t-1}u_{t-1}')$. Additionally, note that $E(u_{t-1}\epsilon_t') = O$. Using these results and applying the vec operator to equation (5.37) yields:

$$\begin{aligned} \text{vec}\,\Sigma_{u,0} &= \text{vec}\,R\Sigma_{u,0}R' + \text{vec}\,\Sigma \\ &= (R \otimes R)\text{vec}\,\Sigma_{u,0} + \text{vec}\,\Sigma \qquad . \end{aligned}$$

This last equation can be solved for vec $\Sigma_{u,0}$:

$$\text{vec } \Sigma_{u,0} = (I_{n^2} - (R \otimes R))^{-1} \text{vec } \Sigma \quad . \tag{5.38}$$

Now, $\Sigma_{u,j}$ will be expressed as a function of R and $\Sigma_{u,0}$. Using the equation $u_t = Ru_{t-1} + \epsilon_t$, postmultiplying the left and right side by u'_{t-j} and taking expectations yields:

$$
\begin{aligned}
\Sigma_{u,j} := E(u_t u'_{t-j}) &= E\left(Ru_{t-1} + \epsilon_t\right) u'_{t-j} \\
&= RE(u_{t-1} u'_{t-j}) + RE(\epsilon_t u'_{t-j}) \\
&= R\Sigma_{u,j-1} \quad .
\end{aligned} \tag{5.39}
$$

By recursive substitution one finds that

$$\Sigma_{u,j} = R^j \Sigma_{u,0} \quad .$$

Since the x_t are non-stochastic, it follows from assumption 5.4 b) that

$$M_j := \lim_{T \to \infty} \sum_{t=j+1}^{T} \frac{x_t x'_{t-j}}{T} \quad .$$

Using proposition 5.4, the definition of M_j and the relation $\Sigma_{u,j} = R^j \Sigma_{u,0}$, the eigenvalues $e_i(F)$ can be expressed as

$$e_i(F) = e_i \left(I_n + \sum_{j=1}^{\infty} \left(\frac{a' M_0^{-1} M_j M_0^{-1} a}{a' M_0^{-1} a} (\Sigma_{u,0}^{-1} R^j \Sigma_{u,0} + R^{j'}) \right) \right) \quad ,$$

where $\Sigma_{u,0}$ can be calculated from equation (5.38).

Example 5.2

In this example, the errors u_t follow a vector MA(1) process and the regressor vector x_t is non-stochastic. The MA(1) error process is defined by

$$u_t = \epsilon_t + \Theta \epsilon_{t-1}$$

with

$$E(\epsilon_t) = 0$$

and

$$E(\epsilon_t \epsilon'_{t-j}) = \begin{cases} \Sigma^* & \text{if } j = 0 \\ O & \text{otherwise} \end{cases} \quad .$$

The variance-covariance matrix of u_t can be calculated as follows (see Hamilton 1994, p.262):

$$\begin{aligned}
\Sigma_{u,0}^* &:= E(u_t u_t') \\
&= E(\epsilon_t + \Theta\epsilon_{t-1})(\epsilon_t + \Theta\epsilon_{t-1})' \\
&= E(\epsilon_t \epsilon_t') + \Theta E(\epsilon_{t-1}\epsilon_{t-1}')\Theta' \\
&= \Sigma^* + \Theta\Sigma^*\Theta' \quad .
\end{aligned} \tag{5.40}$$

The autocovariance matrix of order one is given by

$$\begin{aligned}
\Sigma_{u,1}^* &:= E(u_t u_{t-1}') \\
&= E(\epsilon_t + \Theta\epsilon_{t-1})(\epsilon_{t-1} + \Theta\epsilon_{t-2})' \\
&= \Theta E(\epsilon_{t-1}\epsilon_{t-1}') \\
&= \Theta\Sigma^* \quad .
\end{aligned} \tag{5.41}$$

The autocovariances of higher order than one are all zero:

$$\Sigma_{u,j}^* := E(u_t u_{t-j}') = O \quad \text{for } j \geq 2 \quad .$$

Using these results and $M_j := \lim_{T \to \infty} \sum_{t=j+1}^T x_t x_{t-j}'/T$, the eigenvalues $e_i(F)$ can be expressed as

$$e_i(F) = e_i \left(I_n + \frac{a' M_0^{-1} M_j M_0^{-1} a}{a' M_0^{-1} a}(\Sigma^* + \Theta\Sigma^*\Theta')^{-1}(\Theta\Sigma^* + \Sigma^*\Theta') \right) \quad .$$

5.5 The robust Wald test

In order to avoid the difficulty of identifying the parametric structure of the dynamic process of the errors, the coefficient matrix B can be estimated by GMM and the variance-covariance matrix of this estimator by some heteroskedasticity and autocorrelation consistent (HAC) estimator. In section 4.2.2, three different HAC estimators have been presented: the quadratic spectral estimator (QS), the prewhitened quadratic spectral estimator (PW-QS) (see Andrews 1991 and Andrews and Monahan 1992) and the parametric HAC estimator proposed in Den Haan and Levin (1997).

These results will be used for constructing a robust Wald test for homogeneity.

Assume that the model equation is given by $y_t = Bx_t + u_t$ and, furthermore, that the following holds:

Assumption 5.5

a) $T^{1/2}(\hat{b} - b) \overset{d}{\to} N(0, \Phi)$,

b) $\hat{\Phi}_{HAC} \overset{p}{\to} \Phi$.

Recall that the homogeneity restriction can expressed as

$$R_H b = 0_{n \times 1} \quad .$$

The robust Wald statistic is defined by

$$W_{HAC} := T(R_H \hat{b})' \left(R_H \hat{\Phi}_{HAC} R_H' \right)^{-1} (R_H \hat{b}) \tag{5.42}$$

where $\hat{\Phi}_{HAC}$ denotes a HAC estimator of Φ. Estimation of Φ, where $\Phi = (M^{-1} \otimes I_n) V (M^{-1} \otimes I_n)$, requires a HAC estimator for V. Assuming that assumption 5.5 holds, it follows that under the null hypothesis of homogeneity,

$$T^{1/2} R_H \hat{b} \xrightarrow{d} N(0, R_H \Phi R_H')$$

and

$$\hat{\Phi}_{HAC} \xrightarrow{p} \Phi$$

and, therefore,

$$W_{HAC} \xrightarrow{d} \chi_n^2 \quad ,$$

where n denotes the degree of freedom.
The asymptotic critical values of the Wald test is denoted by χ_α and satisfy

$$P(\chi_n^2 \leq \chi_\alpha) = 1 - \alpha \quad .$$

It is well known that the asymptotic distribution of the Wald statistic can be a very poor approximation of its small sample distribution. A small sample corrected version of this Wald test will, therefore, be specified. Here we propose an "F-version" of the Wald test. This test will be denoted by WF_{HAC}. The motivation for using this test is that the simulation results suggest that the distribution of the W_{HAC} statistic is skewed to the right compared to the chi-square distribution. It is hoped that this effect can be offset by using the approximate relation between F and Wald statistics. Specifically, the WF_{HAC} statistic is defined as

$$WF_{HAC} := \frac{1}{n} \left(\frac{T - 2n - 2}{(T - (n+3))} \right) W_{HAC} \quad . \tag{5.43}$$

The critical values used are based on the F-distribution and satisfy

$$Pr(F_{n;T-2n-2} \leq F_\alpha) = 1 - \alpha \quad ,$$

where F_α denotes the critical value.
The heuristically chosen correction factor $(T - 2n - 2)/(T - (n+3))n$ and the choice of F_α is related to the LRU test. As noted in section 5.2, the LRU test can be interpreted as the F-version of the Wald test.

Invariance of the robust Wald test:
For estimation, one equation of the complete system will be deleted due to adding-up. It should be noted that the robust Wald test defined above is not invariant to the equation deleted. Ray, Ravikumar and Savin (1998, theorem

3) have shown algebraically that the robust Wald test using a HAC estimator of the form $\hat{V}_{HAC} := \hat{\Gamma}_0 + \sum_{j=1}^m k(j,m)(\hat{\Gamma}_j + \hat{\Gamma}_j')$, where the weights $k(j,m)$ are fixed, is invariant to the equation deleted. The result of Ray et al. (1998) applies to the QS estimator if the weights $k(j,m)$ are fixed. The data-dependent weights (bandwidths) are, however, not invariant to the equation deleted. The invariance result does not hold, therefore. The importance of this invariance property can be checked in empirical applications. In the application here, this importance was found to be small and was, therefore, neglected.

5.6 Summary

In this chapter, various tests for homogeneity were discussed under dynamic misspecification and under correct specification.

The transformed likelihood ratio test for the case of normal, homoskedastic errors which are independent in time was denoted by LRU. This test corresponds to Anderson's $\mathcal{U}$ test (see Anderson 1984, 298ff).

It was shown that Anderson's $\mathcal{U}$ statistic under dynamic misspecification is approximately distributed as a quadratic form in normal variables and its distribution can be calculated by Imhof's formula.

When the errors follow a vector autoregressive process, the likelihood ratio statistic (LR) for testing homogeneity was defined. Since only the asymptotic distribution of the LR statistic is known, a small sample corrected likelihood ratio test (LRC) and a Monte Carlo test related to the LR test was proposed (LR-MC).

Finally, a robust Wald test was defined. This Wald test is based on the quasi-maximum likelihood estimation of the model parameters and on the heteroskedasticity and autocorrelation consistent estimation of its variance-covariance matrix. This test was denoted by W_{HAC}. A small sample corrected version of this Wald was, furthermore, proposed (WF_{HAC}).

The following table gives a brief overview over the tests discussed in this chapter:

Table 5.1: Definitions of the homogeneity tests

Statistic	defined in chapter 5	nom. crit. value[1]	assumption
LRU	eq. (5.7), section 5.1, p.50	F_α	$u_t \sim i.i.d.N(O, \Sigma)$
LR	eq. (5.20), sect. 5.3.1, p.56	χ_α	$u_t = \sum_{j=1}^p Ru_{t-j} + \epsilon_t,$ $\epsilon_t \sim i.i.d.N(O, \Sigma)$
LRC	eq. (5.21), sect. 5.3.2, p.57	F_α	$u_t = \sum_{j=1}^p Ru_{t-j} + \epsilon_t,$ $\epsilon_t \sim i.i.d.N(O, \Sigma)$
$LR - MC$	eq. (5.20), sect. 5.3.1, p.60	LR_α^*	$u_t = \sum_{j=1}^p Ru_{t-j} + \epsilon_t,$ $\epsilon_t \sim i.i.d.N(O, \Sigma)$
$^2W_{HAC}$	eq. (5.42), sect. 5.4, p.72	χ_α	see assumption 5.5, p.71
$^3WF_{HAC}$	eq. (5.43), sect. 5.4, p.72	F_α	see assumption 5.5, p.71

[1] F_α satisfies $P(F_{n;T-2n-2} \leq F_\alpha) = 1 - \alpha$ and χ_α satisfies $P(\chi_n^2 \leq \chi_\alpha) = 1 - \alpha$. The critical value LR_α^* is estimated by simulation.

6. Monte Carlo experimentation

In order to analyse by simulation the performance of the various tests for homogeneity discussed in chapter 5, the data-generating process must be defined. As mentioned before, testing for homogeneity can be interpreted as an omitted variable test. The simulation analysis here can be seen as an investigation of the omitted variable test for a linear multivariate regression system with dynamic errors and a specific data-generating process (DGP). The small sample properties of the various test statistics for homogeneity discussed in chapter 5depend on the coefficient matrix B, the regressors x_t and the DGP of the error process. In the simulation study, the sample size and the number of equations will be varied.

Clearly, varying all the parameters in the simulation study would be extremely time-consuming and produce a huge output which could not be interpreted reasonably. The simulation analysis is, therefore, restricted to a small set of DGP's. In order to define a "realistic" DGP, the population parameters will be based on the Rotterdam model estimated with annual and quarterly data from the UK.

The chapter is organised as follows: The data are described in section 6.1. In section 6.2, different DGP's are defined and some computational aspects of the data generation are mentioned. In section 6.3, the experiments are motivated and specified. The results of the simulation study are presented in section 6.4. and section 6.5 concludes.

6.1 Data

In this section, the annual and seasonal data used for estimating the Rotterdam model are described. The definitions of "realistic" DGP's are based on the Rotterdam model and on the estimated population parameters.

Annual data

Annual consumers' nominal and real expenditure data for the UK are taken from "Economic Trends" (Annual Supplement 1996/97) published by the Central Statistical Office. Observations are taken from 1952 to 1995. Consumers' expenditure is given for the following categories: durable goods, food, alcoholic drink and tobacco, clothing and footwear, energy products, other goods, housing (rent, rates and water charges) and other services. Durable goods, housing and other services are excluded. Theoretically, demand for durable goods and services is usually treated within investment theory[1]. This leaves us with five broad categories of commodities. These categories are defined by the "commodity classification" and described in the "United Kingdom National Accounts: Sources and classification" (1985).

In detail, the commodity groups include the following goods:

1. **Food:** bread, cakes and biscuits, other cereals, meat and bacon, fish, milk, cheese and eggs, oils and fats, fruit, potatoes, vegetables, sugar, confectionery, coffee, tea and cocoa, soft drinks, other manufactured food.
2. **Alcoholic drink and tobacco:** beer, spirits, wine, cider and perry, cigarettes, other tobacco.
3. **Clothing and footwear:** men's and boy's wear, women's, girls' and infants' wear, footwear.
4. **Energy:** electricity, gas, coal and coke, other, petrol and oil for transportation.
5. **Other goods**

Population data are taken from the statistical yearbooks. Population data for 1952-1960 are taken from the statistical yearbook from 1977, for 1961-62 from the yearbook 1991 and for 1963-1995 from the yearbook 1996. As there were some minor revisions of the data, the data used correspond to the latest unchanged figures[2].

Total expenditure at period t is defined by the sum of per capita expenditures for the five categories of food, alcoholic drink/tobacco, clothing/footwear, energy and other goods.

The *price index* for one commodity group is calculated by dividing its nominal by its real expenditure at each period.

Levels of aggregation based on annual data: The number of equations will be varied in the simulation study. Here the number of commodity groups ranges between 2 and 5 which means that the number of equations ranges between 1 and 4 since one equation is deleted due to adding-up.

The following definitions of broad commodity groups have been used:

Aggregation level 1: 1. food, 2. miscellaneous: alcoholic drink/tobacco, clothing/footwear, energy products, other goods.

Aggregation level 2: 1. food, 2. alcoholic drink/tobacco, 3. miscellaneous: clothing/footwear, energy products, other goods.

Aggregation level 3: 1. food, 2. alcoholic drink/tobacco, 3. clothing/ footwear, 4. miscellaneous: energy products, other goods.

---

[1] Nevertheless, estimating demand equations with durable goods and services included, give reasonable results if the dynamics is modelled appropriately (see e.g. Deschamps 1993).

[2] Last update: 1997.

Aggregation level 4: 1. food, 2. alcoholic drink/tobacco, 3. clothing/ footwear, 4. energy products, 5. other goods.

The aggregation of the commodity groups is somewhat arbitrary. In order to aggregate consistently, hypotheses have to be made about preferences such as weak separability between the commodity groups (see Deaton and Muellbauer 1980, p.122ff).

Here the purpose of estimating the demand system is simply to obtain some realistic numerical values of the parameters. They will be used to generate the data in order to investigate the properties of the homogeneity test by simulation. We do not focus on the theoretical and empirical aspects of aggregation, therefore.

Seasonal data

Seasonal consumers' nominal and real expenditure data for the UK are again taken from "Economic Trends" (Annual Supplement 1996/97). Observations are taken from 1952 to 1995. Consumers' expenditure for non-durable goods is given for the following four categories: food, clothing and footwear, energy products and other goods. These four broad categories of commodities are defined as above. Seasonal data for alcoholic drink and tobacco are only available since 1970 and are, therefore, excluded.

Seasonal *population data* are not available for the period under consideration. Annual population data are, therefore, extrapolated linearly.

Total expenditure at period t will be defined by the sum of per capita expenditures for the four categories food, clothing/footwear, energy and other goods.

The *price index* for one commodity group is calculated by dividing its nominal by its real expenditure at each period.

Levels of aggregation based on seasonal data: The number of equations will be varied in the simulation study. Here the number of commodities ranges between 2 and 4 which means that the number of equations ranges between 1 and 3 since one equation is deleted due to adding-up.

The following definitions of broad commodity groups have been used:

Aggregation level 1: 1. food 2. miscellaneous: clothing/footwear, energy products, other goods.

Aggregation level 2: 1. food, 2. clothing/footwear, 3. miscellaneous: energy products, other goods.

Aggregation level 3: 1. food, 2. clothing/footwear, 3. energy products, 4. other goods.

6.2 The data-generating process

The DGP is characterised by the model, the distributions of the stochastic variables and the numerical values of the population parameters. Here various DGP's will be defined in terms of the Rotterdam model, as:

$$y_t = Bx_t + u_t$$

with u_t being normally distributed. Homogeneity is imposed for all DGP's because the data will be generated under the null hypothesis of homogeneity.

The definition of the DGP's and the choice of the values of the population parameters, as well as some computational aspects of the data generation, will be described next.

6.2.1 Definition of the data-generating processes

Here five DGP's are defined. Four DGP's should be interpreted as "realistic" DGP's related to the annual data of the Rotterdam model and the other DGP is related to the seasonal data. In the first four DGP's, the errors are generated by a VAR(1) or MA(1) process. Furthermore, the regressors are determined in two ways: they take fixed values or are generated by a VAR(1) process. The DGP's are first described in detail and then summarised for clarity.

DGP1: It is assumed that the errors u_t are generated by a gaussian VAR(1) process given by

$$u_t = Ru_{t-1} + \epsilon_t \quad \text{with} \quad \epsilon_t \sim i.i.d.N(O, \Sigma) \quad .$$

The regressors x_t are fixed and based on the observed annual data described in section 6.1. Restricting the maximal order of the VAR process of u_t to one is not restrictive if working with differenced annual data (see Deschamps 1993). It should be emphasised that here it is not intended to do a general model selection search, but the emphasis is on justifying the data-generating process used in the simulation study. The specific values of (B, R, Σ) for given n will be based on estimation. The Rotterdam model will be estimated by maximum likelihood with homogeneity imposed based on the annual data for the UK. Maximum likelihood estimation is described in Deschamps (1993). The numerical values of (B, R, Σ) for given n are given in the appendices C.1 and C.2.

DGP1*: Instead of taking fixed regressors x_t as for the DGP1 defined above, the regressors are generated randomly. The motivation for using fixed regressors is that, theoretically, it is not clear how the DGP for the $x_t's$ should be defined here. For example generating the x_t from a normal or log-normal distribution can have a great impact on the results (see Davidson and MacKinnon 1993, p.742). Since we are interested in varying the sample size over a large range in the simulation study, a DGP for the x_t is, nevertheless, proposed. The simulation results with fixed and stochastic regressors x_t are compared. In most of our simulation experiments, the results with fixed regressors are similar to those with stochastic regressors.

DGP1* is defined as DGP1 with the fixed regressors replaced by the following DGP for x_t:
Partition x_t such that

$$x_t = \begin{pmatrix} 1 \\ \overline{x}_t \end{pmatrix} . \tag{6.1}$$

It is assumed that the vector $\overline{x}_t$ follows a stationary VAR(1) process:

$$\overline{x}_t - \mu_x = R_x(\overline{x}_{t-1} - \mu_x) + \varsigma_t \quad \text{with } \varsigma_t \sim i.i.d.N(O, \Sigma_\varsigma) , \tag{6.2}$$

where $\mu_x := E(\overline{x}_t)$.

The vector μ_x is estimated by the sample mean of $\overline{x}_t$. The choice of an order of one for the VAR process and the values of (R_x, Σ_ς) are based on the estimation of a VAR model for $(\overline{x}_t - \mu_x)$. The values of $\overline{x}_t$ are related to the annual data described in section 6.1. Recall that the values and dimension of $\overline{x}_t$ depend on the aggregation level used. Here four aggregation levels are considered. Therefore, four VAR models for $\overline{x}_t$ are estimated. In order to justify the order of the VAR model for $\overline{x}_t$, the BIC model selection criterion is used. The estimation was done with the program Microfit 4.0 (Pesaran 1997). The preferred model is a VAR(1) model for $n = 1, 2, 3, 4$. The numerical values of $(\mu_x, R_x, \Sigma_\varsigma)$ for given n are reported in appendix C.4.

DGP2: The errors are now generated by a vector MA(1) process:

$$u_t = \epsilon_t + \Theta\epsilon_{t-1} \quad \text{with } \epsilon_t \sim N(O, \Sigma^*) .$$

The numerical values of the population parameters B and the fixed regressors x_t are the same as for the DGP1 above. The numerical values of (Θ, Σ^*) are chosen such that the variance-covariance and first autocovariance matrix of u_t are approximately the same as those for the VAR(1) error process defined for DGP1. The algorithm to find these values is given in appendix D. The numerical values of (Θ, Σ^*) for given n are reported in appendix C.3.

DGP2*: This DGP is defined as DGP2 with the fixed regressors replaced by the DGP for the x_t defined in equations (6.1) and (6.2).

DGP3: This DGP is related to the seasonal Rotterdam model. It is assumed that the errors u_t are generated by a gaussian VAR(4) process given by

$$u_t = R_1 u_{t-1} + R_2 u_{t-2} + R_3 u_{t-3} + R_4 u_{t-4} + \epsilon_t \quad \text{with} \quad \epsilon_t \sim i.i.d.N(O, \Sigma^{**}) .$$

Seasonal modelling is very difficult in empirical economics. A motivation for specifying the error process in this way is given in Deschamps (1993).

The numerical values of the population parameters $(B, R_1, R_2, R_3, R_4, \Sigma^{**})$ for given n and fixed regressors x_t are based on the maximum likelihood estimators for the seasonal British data with homogeneity imposed (see Deschamps 1993). These numerical values for given n are reported in appendix C.5.

The five DGP's defined above are summarised in the following table:

Table 6.1: Definitions of the DGP's

Model equation: $y_t = Bx_t + u_t$

	DGP of the errors	DGP of the regressors
DGP1	$u_t = Ru_{t-1} + \epsilon_t,$ $\epsilon_t \sim i.i.d.N(O, \Sigma)$	fixed (based on annual data)
DGP1*	$u_t = Ru_{t-1} + \epsilon_t,$ $\epsilon_t \sim i.i.d.N(O, \Sigma)$	$\overline{x}_t - \mu_x = R_x(\overline{x}_{t-1} - \mu_x) + \varsigma_t,$ $\varsigma_t \sim i.i.d.N(O, \Sigma_\varsigma)$
DGP2	$u_t = \epsilon_t + \Theta\epsilon_{t-1},$ $\epsilon_t \sim i.i.d.N(O, \Sigma^*)$	fixed (based on annual data)
DGP2*	$u_t = \epsilon_t + \Theta\epsilon_{t-1},$ $\epsilon_t \sim i.i.d.N(O, \Sigma^*)$	$\overline{x}_t - \mu_x = R_x(\overline{x}_{t-1} - \mu_x) + \varsigma_t,$ $\varsigma_t \sim i.i.d.N(O, \Sigma_\varsigma)$
DGP3	$u_t = \sum_{j=1}^{4} R_j u_{t-j} + \epsilon_t,$ $\epsilon_t \sim i.i.d.N(O, \Sigma^{**})$	fixed (based on seasonal data)

The values of the population parameters can be found in appendix C:

– B for DGP1, DGP1*, DGP2, DGP2* in appendix C.1,
– (R, Σ) for DGP1, DGP1* in appendix C.2,
– (Θ, Σ^*) For DGP2, DGP2* in appendix C.3,
– $(\mu_x, R_x, \Sigma_\varsigma)$ for DGP1* and DGP2* in appendix C.4,
– $(B, R_1, R_2, R_3, R_4, \Sigma^{**})$ for DGP3 in appendix C.5.

6.2.2 Computational aspects

In this subsection, we give some details of how observations from a VAR(1), VAR(4) and MA(1) processes can be generated. The formulas for calculating the autocovariances are given, furthermore. The latter will be used for approximating the distribution of the LRU statistic based on the asymptotic result discussed in section 5.4.

Generation of observations from a VAR(1) process: For DGP1 and DGP1* it is assumed that the errors u_t are generated by a gaussian VAR(1) process given by

$$u_t = Ru_{t-1} + \epsilon_t \quad \text{with} \quad \epsilon_t \sim i.i.d.N(O, \Sigma) \quad .$$

Generating u_t requires generating ϵ_t from a $i.i.d.N(O, \Sigma)$ distribution and the values of the initial vector u_0. The error vector ϵ_t can be generated as follows: the Cholesky factorisation of Σ is of the form $\Sigma = A'A$. Furthermore, note that $A'N(O, I) = N(O, A'A) = N(O, \Sigma)$. Therefore, for generating the ϵ_t one first generates a random vector from a $N(O, I)$ distribution (which is a standard command in many computer programs) and premultiplies this vector by A'.

The values of u_t are calculated from the recursive relation $u_t = Ru_{t-1} + \epsilon_t$ for $t = 1, 2, \ldots, T$. The initial vector u_0 will be drawn from a multivariate normal distribution with

$$u_0 \sim N(0, \Sigma_{u,0}) \quad,$$

where $\Sigma_{u,0} := E(u_t u_t')$.

The variance-covariance and the autocovariance matrices of u_t can be calculated as shown in example 5.1 in section 5.4. There it was shown that for given (R, Σ), the values of $\Sigma_{u,0}$ can be calculated from the equation

$$\text{vec } \Sigma_{u,0} = (I_{n^2} - (R \otimes R))^{-1} \text{ vec } \Sigma \quad.$$

The autocovariance matrix $\Sigma_{u,j} := E(u_t u_{t-j}')$ is given by

$$\Sigma_{u,j} = R^j \Sigma_{u,0} \quad.$$

Generation of observations from a VAR(4) process: For DGP3 it is assumed that the errors u_t are generated by a gaussian VAR(4) process given by

$$u_t = \sum_{j=1}^{4} R_j u_{t-j} + \epsilon_t \quad \text{with} \quad \epsilon_t \sim i.i.d. N(O, \Sigma^{**}) \quad.$$

The values of u_t are calculated from the recursive relation $u_t = \sum_{j=1}^{4} R_j u_{t-1} + \epsilon_t$ for $t = 1, 2, \ldots, T$. The initial vectors $u_{-3}, u_{-2}, u_{-1}, u_0$ will be generated as follows:
Let

$$\xi_t := \begin{pmatrix} u_{t-3} \\ u_{t-2} \\ u_{t-1} \\ u_t \end{pmatrix} \quad.$$

These initials vectors are drawn from a multivariate distribution with

$$\xi_0 \sim N(O, \Sigma_{\xi,0}) \quad,$$

where $\Sigma_{\xi,0} := E(\xi_t \xi_t')$.

The variance-covariance matrix $\Sigma_{\xi,0}$ is calculated as follows: note that ξ_t has a VAR(1) representation:

$$\xi_t := \begin{pmatrix} u_{t-3} \\ u_{t-2} \\ u_{t-1} \\ u_t \end{pmatrix} = \begin{pmatrix} O_{n \times n} & I_n & O_{n \times n} & O_{n \times n} \\ O_{n \times n} & O_{n \times n} & I_n & O_{n \times n} \\ O_{n \times n} & O_{n \times n} & O_{n \times n} & I_n \\ R_4 & R_3 & R_2 & R_1 \end{pmatrix} \begin{pmatrix} u_{t-4} \\ u_{t-3} \\ u_{t-2} \\ u_{t-1} \end{pmatrix} + \begin{pmatrix} O_{n \times 1} \\ O_{n \times 1} \\ O_{n \times 1} \\ \epsilon_t \end{pmatrix}$$

$$(6.3)$$

which can be written compactly as

$$\xi_t = F\xi_{t-1} + v_t \quad ,$$

where the definitions of F and v_t follow from equation (6.3). The procedure of finding the variance-covariance matrix of a VAR(1) process has been described above. Therefore,

$$\text{vec } \Sigma_{\xi,0} = (I_{(4n)^2} - (F \otimes F))^{-1}\text{vec } \Sigma_v \quad ,$$

with

$$\Sigma_v := E(v_t v_t') = \begin{pmatrix} O_{n \times n} & O_{n \times n} & O_{n \times n} & O_{n \times n} \\ O_{n \times n} & O_{n \times n} & O_{n \times n} & O_{n \times n} \\ O_{n \times n} & O_{n \times n} & O_{n \times n} & O_{n \times n} \\ O_{n \times n} & O_{n \times n} & O_{n \times n} & \Sigma^{**} \end{pmatrix} \quad .$$

The autocovariance matrix $\Sigma_{u,j}^{**} := E(u_t u_{t-j}')$, where u_t is VAR(4), can be calculated as follows: note that the first n rows and columns of $\Sigma_{\xi,j} := E(\xi_t \xi_{t-j}')$ equal $\Sigma_{u,j}^{**}$. To find $\Sigma_{u,j}^{**}$, the autocovariance matrix $\Sigma_{\xi,j}$ is, therefore, calculated first. Since ξ_t follows a VAR(1) process, the autocovariance matrix $\Sigma_{\xi,j}$ can be calculated as

$$\Sigma_{\xi,j} = F^j \Sigma_{\xi,0} \quad .$$

Generation of observations from a vector MA(1) process: For DGP2 and DGP2* it is assumed that the errors u_t are generated by a vector MA(1) process. The values of u_t are calculated from the recursive relation $u_t = \epsilon_t + \Theta\epsilon_{t-1}$ for $t = 1, 2, \ldots, T$. The initial vector ϵ_0 will be drawn from a multivariate normal distribution with

$$\epsilon_t \sim N(O, \Sigma^*) \quad .$$

The variance-covariance matrix $\Sigma_{u,j}^* := E(u_t u_t')$ is given by (see section 5.4, example 5.2)

$$\Sigma_{u,0}^* = \Sigma^* + \Theta\Sigma^*\Theta' \quad .$$

The autocovariance matrix $\Sigma_{u,j}^* := E(u_t u_{t-j}')$ is given by

$$\Sigma_{u,j}^* = \begin{cases} \Theta\Sigma^* & \text{for } j = 1 \\ O & \text{for } j \geq 2 \end{cases} \quad .$$

6.3 Experiments

The simulation experiments on the various homogeneity tests are divided into three main parts: the parametric test under dynamic misspecification (LRU test), the dynamically correctly specified likelihood ratio test (LR) and its related versions (LRC, LR-MC) and the various versions of the robust Wald test W_{HAC} discussed in section 5.4. This subsection is organised as follows:

In section 6.3.1, the simulation estimator of the test's type I error for given DGP is described. Section 6.3.2 motivates and specifies the experimental design for the LRU test, section 6.3.3 for the LR, LRC and LR-MC tests and section 6.3.4 for the robust Wald test. For clarity, the experimental design is summarised in section 6.3.5.

6.3.1 Estimating the test's true type I error by simulation

For the various tests for homogeneity, the type I error will be estimated for given nominal critical value and DGP under the null hypothesis of homogeneity. The rejection frequency is estimated as follows (see Davidson and MacKinnon 1993, p.739): at each replication of the experiment, the null hypothesis of homogeneity is rejected if the observed statistic exceeds the nominal critical value. Since a realisation of the test statistic is generated independently at each replication, the experiment can be seen as N independent Bernoulli trials, where N denotes the number of replications. The estimated rejection frequency denoted by $\bar{p}$ is calculated by dividing the number of rejections by the number of replications. Let t_j be the jth generated test statistic and t_α be the nominal critical value. Then $\bar{p}$ is estimated by

$$\bar{p} = N^{-1} \sum_{j=1}^{N} I(t_j > t_\alpha) \quad,$$

where the indicator function $I(\cdot)$ takes the value one if its argument is positive and zero otherwise.

Note that the estimated rejection frequency is distributed as a binomial. The variance of this estimator is given by $\bar{p}(1 - \bar{p})/N$. In order to evaluate the quality of the simulation estimator, the length of its 95% interval can be estimated. Using the fact that for large N, the standard normal distribution is a good approximation of the binomial, the length of the confidence interval can be estimated by

$$2 \cdot 1.96 \left(\frac{\bar{p}(1 - \bar{p})}{N} \right)^{1/2} \quad,$$

where 1.96 is the critical value of the standard normal distribution. Note that the maximal estimated length of the confidence interval for given N is achieved if $\bar{p} = 0.5$ and the minimum if $\bar{p} = 0$. In the simulation study, the number of replications N equals 10000 or 20000 (except for the LR-MC test). Therefore, the estimated length of the confidence interval for $N = 10000$ is bounded by that of the interval $[0; 0.020]$ and for $N = 20000$ by that of $[0; 0.014]$.

6.3.2 Bias of the misspecified LRU test

In section 5.1, the LRU test has been developed for testing homogeneity. Recall from section 5.1 that the LRU test is unbiased if the errors are normal, homoskedastic and independent in time with the LRU statistic being distributed as an $F_{n;T-2n-2}$ under the null hypothesis.

Here we are interested in studying the importance and direction of the bias of the LRU test under *dynamic misspecification*. By the definition of the DGP's, the dynamic misspecification is due to the time-dependence of the errors. It is interesting to see how the bias of this test is related to the number of equations used. The rejection frequencies of the LRU test are estimated by simulation for the five DGP's described above. These results are, furthermore, compared with the approximate asymptotic rejection frequencies of the LRU test. The asymptotic approximation is based on Imhof's formula (see section 5.4, p.65 and p.67, equation (5.35)).

Rejection frequency of the misspecified LRU statistic: At each replication of the experiment, one observation of the LRU statistic can be calculated as described in section 5.1. The nominal critical value F_α satisfies the relation

$$P(F_{n,T-2n-2} \leq F_\alpha) = 1 - \alpha \quad .$$

Recall that the nominal type I error of the LRU test would be identical to the true type I error of the test if the errors were normal, homoskedastic and independent in time. For given (T, n) and DGP, each experiment is replicated 20000 times.

Approximating the distribution of the LRU statistic by numerical integration: In section 5.4, the asymptotic distribution of the LRU statistic was derived. There it was proposed to approximate the finite sample null distribution of the LRU statistic by

$$\Pr(LRU \leq F_\alpha) \approx \Pr(\sum_{i=1}^{n} \lambda_i z_i^2 \leq \chi_\alpha) \quad ,$$

where F_α satisfies $P(F_{n;T-2n-2} \leq F_\alpha) = 1 - \alpha$ and χ_α satisfies $P(\chi_n^2 \leq \chi_\alpha) = 1 - \alpha$. The z_i^2's are independent central chi-square random variables with one degree of freedom.

The λ_i's are defined by (see proposition 4.4)

$$\lambda_i = e_i \left(I_n + \sum_{j=1}^{\infty} \frac{a' M_0^{-1} M_j M_0^{-1} a}{a' M_0^{-1} a} \Sigma_{u,0}^{-1}(\Sigma_{u,j} + \Sigma_{u,j}') \right) \quad \text{for } i = 1, 2, \ldots, n,$$

where

$$\Sigma_{u,j} := E(u_t u_{t-j}') \quad \forall j, t$$

and

$$M_j := \lim_{T \to \infty} E\left(\sum_{t=j+1}^{T} \frac{x_t x'_{t-j}}{T} \right) \quad .$$

Here the interest is in seeing how close the small sample distribution of the LRU statistic is to its approximate distribution. The approximate rejection frequency is calculated by numerical integration based on Imhof's formula. The examples of the asymptotic distribution of the LRU statistic presented in section 5.4 are related to the DGP1 and DGP2. For each of the five DGP's defined above we briefly point out how the eigenvalues λ_i are calculated:

Calculation of the λ_i for DGP1: Here the errors are VAR(1) and the regressors x_t are fixed. This DGP corresponds to example 5.1 in section 5.4. There it was shown that

$$\lambda_i = e_i \left(I_n + \sum_{j=1}^{\infty} \frac{a' M_0^{-1} M_j M_0^{-1} a}{a' M_0^{-1} a} (\Sigma_{u,0}^{-1} R^j \Sigma_{u,0} + R^{j'}) \right) \quad \text{for } i = 1, 2, \ldots, n.$$

Since the number of observations is limited, M_j is approximated by $\sum_{t=j+1}^{T} x_t x'_{t-j}/T$ for $j = 0, 1, 2, \ldots, T$ and M_j is set to zero for $j > T$. For the specific values of x_t used here, the elements of M_j are close to zero for j "large enough". Therefore, the approximation error of ignoring M_j for $j > T$ is considered to be small.

Calculation of the λ_i for DGP1*: The errors are VAR(1) as above. The regressors x_t without their constant term also follow a stationary VAR(1) process. The eigenvalues λ_i can be calculated as for DGP1 with the values of M_j replaced as follows: the stationarity of x_t implies that

$$M_j := \lim_{T \to \infty} E\left(\sum_{t=j+1}^{T} \frac{x_t x'_{t-j}}{T} \right) = E(x_t x'_{t-j}) \quad .$$

For calculating M_j recall that

$$x_t = \begin{pmatrix} 1 \\ \overline{x}_t \end{pmatrix}$$

and

$$\overline{x}_t - \mu_x = R_x(\overline{x}_{t-1} - \mu_x) + \varsigma_t \quad \text{with} \quad \varsigma_t \sim i.i.d. N(0, \Sigma_\varsigma) \quad ,$$

where $\mu_x := E(\overline{x}_t)$.
Now,

$$M_j = E(x_t x'_{t-j}) = E\left(\begin{pmatrix} 1 \\ \overline{x}_t \end{pmatrix} (1 \quad \overline{x}'_{t-j}) \right) = \begin{pmatrix} 1 & \mu'_x \\ \mu_x & E(\overline{x}_t \overline{x}'_{t-j}) \end{pmatrix} \quad .$$

The term $E(\overline{x}_t \overline{x}'_{t-j})$ can be calculated as follows: defining

$$\Sigma_{\overline{x},j} := E(\overline{x}_t - \mu_x)(\overline{x}_{t-j} - \mu_x)'$$

and solving for $E(\overline{x}_t \overline{x}'_{t-j})$ yields

$$E(\overline{x}_t \overline{x}'_{t-j}) = \Sigma_{\overline{x},j} + \mu_x \mu'_x \quad .$$

Note that $(\overline{x}_t - \mu_x)$ is a zero mean stationary VAR(1) process. Using the same arguments as in example 5.1 in section 5.4, it can be shown that

$$\Sigma_{\overline{x},j} = R_x^j \Sigma_{\overline{x},0} \quad .$$

Therefore,

$$E(\overline{x}_t \overline{x}'_{t-j}) = R_x^j \Sigma_{\overline{x},0} + \mu_x \mu'_x \quad .$$

The matrix $\Sigma_{\overline{x},0}$ can be calculated from

$$\text{vec}\, \Sigma_{\overline{x},0} = \left(I_{(n+2)^2} - (R_x \otimes R_x)\right)^{-1} \text{vec}\, \Sigma_\varsigma \quad .$$

Calculation of the λ_i for DGP2: Here the errors are MA(1) and the regressors x_t are fixed. This DGP corresponds to example 5.2 in section 5.4. There it was shown that

$$\lambda_i = e_i \left(I_n + \frac{a' M_0^{-1} M_j M_0^{-1} a}{a' M_0^{-1} a}(\Sigma^* + \Theta \Sigma^* \Theta')^{-1}(\Theta \Sigma^* + \Sigma^* \Theta')\right)$$

for $i = 1, 2, \ldots, n$. The M_j are approximated as described for DGP1.

Calculation of the λ_i for DGP2*: Here the errors are also MA(1) but the regressors x_t without their constant term follow a stationary VAR(1) process. The calculation of the λ_i's is the same as for DGP2 if the M_j are calculated as for DGP1*.

Calculation of the λ_i for DGP3: Here the errors are VAR(4) and the regressors are fixed. The calculation of the λ_i's is based on the general formula

$$\lambda_i = e_i \left(I_n + \sum_{j=1}^{\infty} \frac{a' M_0^{-1} M_j M_0^{-1} a}{a' M_0^{-1} a} \Sigma_{u,0}^{**-1}(\Sigma_{u,j}^{**} + \Sigma_{u,j}^{**'})\right) \quad \text{for } i = 1, 2, \ldots, n.$$

The matrix M_j is approximated by $\sum_{t=j+1}^{T} x_t x'_{t-j}/T$ for $j = 0, 1, 2, \ldots, T$ and M_j is set to zero for $j > T$. The formula for calculating the $\Sigma_{u,j}^{**} := E(u_t u'_{t-j})$, where u_t is VAR(4), is described in section 6.2.2.

For given nominal critical value, the rejection frequency can be approximated by numerical integration based on Imhof's formula (see section 5.4, equation (5.29)). Numerical integration was done with the program GAUSS (version 3.2.12). Problems occurred if the upper integration limit was set too large. The function to integrate oscillates but its amplitudes decays. Graphical analysis suggests how the truncation point of integration should be chosen.

The exact results for the numerical problem given in Imhof (1961, table 1, first example, p.424) could be reproduced successfully[3].

The values of the λ_i's for the different DGP's are reported in the following table:

Table 6.2: Numerical values of the λ_i

DGP	n	λ_i
DGP1	1	$\lambda_1 = 1.4065$
	2	$\lambda_1 = 2.0144, \lambda_2 = 0.6271$
	3	$\lambda_1 = 1.8416, \lambda_2 = 1.6557, \lambda_3 = 0.6828$
	4	$\lambda_1 = 2.2278, \lambda_2 = 1.7852, \lambda_3 = 1.1045, \lambda_4 = 0.5198$
DGP1*	1	$\lambda_1 = 1.4347$
	2	$\lambda_1 = 2.1596, \lambda_2 = 0.6227$
	3	$\lambda_1 = 1.9261, \lambda_2 = 1.7048, \lambda_3 = 0.6841$
	4	$\lambda_1 = 2.5042, \lambda_2 = 1.8376, \lambda_3 = 1.0899, \lambda_4 = 0.5164$
DGP2	1	$\lambda_1 = 1.3363$
	2	$\lambda_1 = 1.6579, \lambda_2 = 0.5300$
	3	$\lambda_1 = 1.5934, \lambda_2 = 1.4960, \lambda_3 = 0.6241$
	4	$\lambda_1 = 1.6812, \lambda_2 = 1.5796, \lambda_3 = 1.1376, \lambda_4 = 0.3197$
DGP2*	1	$\lambda_1 = 1.3467$
	2	$\lambda_1 = 1.6832, \lambda_2 = 0.5119$
	3	$\lambda_1 = 1.5989, \lambda_2 = 1.5006, \lambda_3 = 0.6207$
	4	$\lambda_1 = 1.7001, \lambda_2 = 1.5956, \lambda_3 = 1.1414, \lambda_4 = 1.3009$
DGP3	1	$\lambda_1 = 0.8703$
	2	$\lambda_1 = 3.4650, \lambda_2 = 0.9616$
	3	$\lambda_1 = 5.8387, \lambda_2 = 2.3502, \lambda_3 = 0.9812$

6.3.3 The small sample bias of the correctly specified likelihood ratio test

This subsection deals with the *correctly specified* likelihood test for homogeneity. In section 5.3.1, the correctly specified likelihood ratio statistic (LR) for testing homogeneity was defined. Under the null hypothesis, the LR statistic is asymptotically distributed as a χ^2 with n degrees of freedom. It is well known that the asymptotic distribution of the LR statistic is often a poor approximation of its small sample distribution. Therefore, a "corrected" likelihood ratio (LRC) test was proposed which is related to Anderson's $\mathcal{U}$ test (see section 5.3.2). Furthermore, a Monte Carlo test was proposed in section 5.3.3.

Here we are interested in studying the small sample performance of the dynamically correctly specified LR test for homogeneity. Furthermore, it will

---

[3] Integration in GAUSS is based on the Gauss-Legendre quadrature algorithm. It is important to set the order of the integration to its maximum value which is 40.

be checked wether if the small sample corrected version of the LR test and the Monte Carlo test are valid for the examples here.

The nominal critical value χ_α of the LR test satisfies the relation

$$P(\chi_n^2 \leq \chi_\alpha) = 1 - \alpha \quad .$$

The nominal critical value F_α of the LRC test satisfies the relation

$$P(F_{n,T-2n-2} \leq F_\alpha) = 1 - \alpha \quad .$$

Each experiment is replicated 10000 times for given (T, n) and DGP.

The Monte Carlo test (LR-MC) consists in estimating the critical value of the LR test by simulation. Specifically, the model is estimated under the null hypothesis of homogeneity for which the LR statistic is calculated. For the ith replication of the experiment, the realised LR statistic is denoted by LR_i. The constrained estimates are used to define a sample-based (artificial) DGP. This DGP will be used to generate N_B artificial data sets. For each of the N_B data sets, the LR statistic is calculated. The jth realisation of the bootstrap-based LR statistic within the ith replication of the experiment is denoted by LR_i^{*j}. One can thus estimate the true distribution of the LR statistic if the data were generated by the *artificial* DGP. This estimated distribution is used as an approximation of the distribution of the LR statistic implied by the *true* DGP. The bootstrapped-based critical value $LR_{\alpha,i}^{*j}$ is chosen such that

$$N_B^{-1} \sum_{j=1}^{N_B} I(LR_i^{*j} > LR_{\alpha,i}^*) = \alpha \quad ,$$

where α denotes the nominal type I error of the test.

For the ith simulation run, the null hypothesis of homogeneity is rejected if

$$LR_i > LR_{\alpha,i}^* \quad .$$

It should be noted that $LR_{\alpha,i}^*$ is a random variable converging to a constant if N_B tends to infinity. This implies that the choice of N_B should not be chosen too small in order to reduce the variability of $LR_{\alpha,i}^*$. Here we set $N_B = 100$ which is often used in applied work. The estimation of the type I error of the Monte Carlo test is computationally very demanding. The true type I error of the LR-MC test is estimated by

$$N^{-1} \sum_{i=1}^{N} I(LR_i > LR_{\alpha,i}^*) \quad .$$

For example, generating $N = 10000$ realisations of the Monte Carlo test with $N_B = 100$ would require that the LR statistic is calculated 10^7 times. Note that the calculation of the LR statistic is already time-consuming. The model must be estimated under the null and alternative hypothesis. The model is nonlinear in the parameters and the estimation algorithm described in Deschamps (1993) is iterative and guarantees that the estimators convergence globally. In order to reduce the computational costs, we proceed as follows: first, the number of replications of the simulation was set to $N = 1000$. Second, the maximal number of iterations for the estimation algorithm was set to 10. The initial estimates correspond to the parameters values of the DGP from which the data are generated. Limiting the number of iterations is not problematic here. It was verified that the distribution of the LR statistic is hardly affected by the procedure used here.

6.3.4 The small sample bias of the robust Wald test

The GMM estimation of the coefficients of the Rotterdam model and the heteroskedasticity and autocorrelation consistent (HAC) estimation of their variance-covariance matrix was motivated by the fact that these estimators are consistent under fairly general conditions. One does, therefore, not have to identify the dynamic structure of the errors, and this is desirable since discriminating between various dynamic forms of the error process is very difficult. In chapter 4, three types of HAC estimators have been described. These were the quadratic spectral (QS) and the prewhitened quadratic spectral (PW-QS) kernel estimators proposed in Andrews (1991) and Andrews and Monahan (1992) and the parametric VARHAC estimator proposed in Den Haan and Levin (1997). The related forms of the Wald tests were described in section 5.5. These Wald tests are robust in the sense that the Wald statistics are asymptotically distributed as a chi-square with n degrees of freedom under quite general conditions.

The motivation of the simulation study is to evaluate the costs of this "robust" approach. Theory does not indicate the direction and the importance of bias for these Wald tests and small sample corrected versions of these tests have been proposed, therefore. The small sample corrected Wald test can be interpreted as the F-version of the Wald test and is denoted by WF_{HAC}. It should be noted that the small sample version of the Wald test has no rigorous justification.

The nominal critical values of the W_{HAC} test is denoted by χ_α and satisfies the relation

$$P(\chi_n^2 \leq \chi_\alpha) = 1 - \alpha \quad .$$

For the WF_{HAC} test, the nominal critical value F_α satisfies

$$P(F_{n,T-2n-2} \leq F_\alpha) = 1 - \alpha \quad .$$

For given (T, n) and DGP, each experiment is replicated 20000 times.

The robust Wald test based on the QS estimator is described in section 5.5. For applying the PW-QS estimator, the prewhitening procedure has to be specified. Recall that $\hat{v}_t$ was defined by $x_t \otimes \hat{u}_t$, which is of the dimension $n(n+3) \times 1$. Here two procedures of prewhitening (filtering) $\hat{v}_t$ are considered.

Unrestricted VAR(1) prewhitening procedure: The first procedure consists in estimating an unrestricted VAR(1) model for $\hat{v}_t$. The QS estimator is applied to the resulting residuals of this regression. The W_{HAC} test based on the unrestricted PW-QS estimator is denoted by W_{PW} and the related F-version by WF_{PW}.

Restricted VAR(1) prewhitening procedure: The second procedure of prewhitening requires estimating a restricted VAR representation of $\hat{v}_t$ based on the BIC selection criterion. The aspects of technical implementation for the simulation study will be specified. A restricted form of the following model is estimated:

$$\hat{v}_t = \sum_{j=1}^{p} A_j \hat{v}_{t-j} + v_t^* \quad ,$$

where the A_j are coefficient matrices of dimension $n(n+3) \times n(n+3)$ which have to be estimated. The QS estimator is applied to the estimated values of v_t^*. Here the maximal order of the VAR model was set to one for DGP1 and DGP1* and to four for DGP3. In order to keep the estimation procedure for the simulation study simple, the estimation of the A_j are based on the least-squares estimation of each component of $\hat{v}_t$ on its lagged values and on the current and lagged values of all the other elements of $\hat{v}_t$. Note that this is not an efficient estimator. In order to choose the preferred restricted form of this univariate regression, the BIC criterion has to be calculated for all possible restricted forms. Restricting means that one or more regressors are deleted. For simplifying the simulation on the computer, we proceed as follows:

We distinguish two sorts of regressors: the lagged regressands and the other regressors. The lag order of the lagged regressands is between zero and the maximal lag order p. The lag order of the other regressors is also between zero and p but restricted to be the same.

The GAUSS procedure of implementing this approach used in this simulation study is based on the GAUSS procedure for the VARHAC estimator written by Den Haan and Levin (some small modifications make this procedure much faster). The W_{HAC} test based on the restricted PW-QS estimator is denoted by W_{RPW} and the related F-version by WF_{RPW}.

For the VARHAC estimator a restricted VAR(p) model for $\hat{v}_t$ has to be estimated. Here the estimation of the restricted VAR model is exactly the same as for the PW-QS estimator if the prewhitening procedure is based on the BIC model selection criterion.

6.3.5 Summary of the experimental design

In this subsection the motivation of the simulation study is discussed and the experimental design summarised.

As discussed in section 5, the LRU test should be applied if the errors are normal, homoskedastic and independent in time. Here the LRU test is studied under *dynamic misspecification*. The simulation experiments on the LRU test are motivated by two questions: first, we are interested in studying the importance and direction of the bias of the LRU test under dynamic misspecification for specific examples. Second, it will be checked how close the small sample distribution of the LRU statistic is to its approximate asymptotic distribution based on Imhof's formula (see section 5.4, p.65 and p.67).

In section 5.3, the correctly specified the likelihood ratio test (LR) statistic for testing homogeneity was defined if the errors follow a vector autoregressive process. Since only the asymptotic distribution of this test is known, a small sample corrected likelihood ratio test (LRC) and a Monte Carlo test was proposed (LR-MC). The motivation of the experiments on these three dynamically correctly specified tests is to study their small sample performance. The relation between the small sample bias and the number of equations n and observations T is analysed in detail.

The robust Wald test is based on the quasi-maximum likelihood estimation of the model parameters and on the heteroskedasticity and autocorrelation consistent estimation of its variance-covariance matrix. This test was denoted by W_{HAC} and its related F-version by WF_{HAC}. The motivation of the simulation experiments on the W_{HAC} test is to evaluate the bias of the various "robust" Wald tests for specific examples. We are interested is studying the relation between the small sample bias and the number of equations n and observations T.

The considered Wald tests are the

- W_{QS} test based on the quadratic spectral HAC estimator proposed in Andrews (1991),
- W_{PW} test based on the prewhitening quadratic spectral HAC estimator proposed in Andrews and Monahan (1992),
- W_{RPW} test based on the restricted prewhitening quadratic spectral HAC estimator discussed in section 6.3.4 and the
- W_{VH} test based on the VARHAC estimator proposed in Den Haan and Levin (1997).

It will also be checked if the related F-versions WF_{QS}, WF_{PW}, WF_{RPW}, WF_{VH} perform better.

The experimental design is summarised in the following table:

Table 6.3: Experimental design

Statistic	defined in ch. 5	nom. crit. value[1]	DGP	variation of n	variation of T
LRU	eq. (5.7), p.50	F_α	DGP1	$1,2,3,4$	$30,36,42$
			DGP1*	$1,2,3,4$	$30,36,42,$ $50,100,250$
			DGP2	$1,2,3,4$	$30,36,42$
			DGP2*	$1,2,3,4$	$30,36,42,$ $50,100,250$
			DGP3	$1,2,3$	$100,130,156$
LR	eq. (5.20), p.56	χ_α	DGP1	$1,2,3,4$	$30,36,42$
			DGP1*	$1,2,3,4$	$30,36,42,$ $50,100,250$
			DGP3	$1,2,3$	$100,130,156$
LRC	eq. (5.21), p.57	F_α	DGP1	$1,2,3,4$	$30,36,42$
			DGP1*	$1,2,3,4$	$30,36,42,$ $50,100,250$
			DGP3	$1,2,3$	$100,130,156$
$LR-MC$	eq. (5.20), p.56	$^2LR_\alpha^*$	DGP1	$1,2,3,4$	42
			DGP1*	$1,2,3,4$	42
			DGP3	$1,2,3$	156
$^3W_{HAC}$	eq. (5.42), p.72	χ_α	DGP1	$1,2,3,4$	$30,36,42$
			DGP1*	$1,2,3,4$	$42,50,100,$ $250,500$
			DGP3	$1,2,3$	$100,130,156$
$^4WF_{HAC}$	eq. (5.43), p.72	F_α	DGP1	$1,2,3,4$	$30,36,42$
			DGP1*	$1,2,3,4$	$42,50,100,$ $250,500$
			DGP3	$1,2,3$	156

[1] F_α satisfies $P(F_{n;T-2n-2} \leq F_\alpha) = 1 - \alpha$ and χ_α satisfies $P(\chi_n^2 \leq \chi_\alpha) = 1 - \alpha$.
[2] The calculation of the bootstrap critical value LR_α^* is described in section 5.3.3, p.60.
[3] The W_{HAC} tests include $W_{QS}, W_{PW}, W_{RPW}, W_{VH}$.
[4] The WF_{HAC} tests include $WF_{QS}, WF_{PW}, WF_{RPW}, WF_{VH}$.

6.4 Simulation results

In this section, the simulation results for the various homogeneity tests are presented. In section 6.4.1, some aspects of the presentation of the simulation results are discussed and section 6.4.2. gives a brief overview of the simulations results. Section 6.4.3 presents the simulation results for the misspecified LRU test, section 6.4.4 for the correctly specified LR, LRC, LR-MC test and section 6.4.5 for the robust Wald tests.

6.4.1 Aspects of the presentation of the simulation results

The small sample bias of a test be characterised by reporting the estimated true type I error of the test for given DGP and its associated nominal type I error α.

For the LRU test, for example, the rejection frequency for given α is estimated by

$$N^{-1} \sum_{j=1}^{N} I(LRU_j > F_\alpha) \quad ,$$

where LRU_j denotes the jth realisation of the LRU statistic. The nominal critical value F_α satisfies

$$P(F_{n;T-2n-2} \leq F_\alpha) = 1 - \alpha \quad ,$$

where α is the nominal type I error of the test.

The simulation results will be presented by reporting numerical results and by graphics. Reporting the numerical simulation results in tables has some drawbacks (see Davidson and MacKinnon, 1998). The interpretation of the reported numerical results is not straightforward for the reader and usually restricted to the nominal significance levels of $0.01, 0.05$ and 0.10. Here we are interested in seeing the importance and direction of the test's bias for a given α. Since all our examples are specific, the exact numerical simulation results are not very interesting as such. We, therefore, only report the numerical estimated true type I errors for given $\alpha = 0.05$ for the most important results. For a more detailed analysis of the simulation results, we use the graphical methods advocated in Davidson and MacKinnon (1998).

The numerical estimated true type I errors for given nominal type I error of 5% are reported in table 6.4a and 6.4b for the most important simulation results. This gives a good overview of the simulation results in a synthesised form. The simulation results for each of the considered tests are discussed below.

In order to analyse these results in detail we will use graphical methods. The *bias* of the test is defined by the estimated true minus the nominal type I error. Two types of graphics are used here. In the first type of graphic, the test's bias is plotted against the nominal type I error. Davidson and

MacKinnon (1998) call this p-value discrepancy plots. By reporting the test's bias, one can easily see how much a specific test over or underrejects the null hypothesis for a large range of α. The nominal size ranges from 0.001 to 0.20. In the second type of graphic, the test's bias for given $\alpha = 0.05$ is plotted against the number of observation T. This plot gives a good impression on how fast the test statistic under consideration converges to its asymptotic distribution.

Table 6.4a: Estimated rejection frequencies in % for given $\alpha = 5\%$

DGP	n	T	LRU	LR	LRC	LR-MC
DGP1*	1	30	9.0	9.7	7.0	-
		36	9.2	9.0	7.0	-
		42	9.5	8.8	7.0	6.3
		50	9.1	7.9	6.5	-
		100	9.8	6.3	5.7	-
		250	10.2	5.4	5.2	-
		500	-	-	-	-
[1] Imhof's approx.			10.2	-	-	
	2	30	9.8	15.1	9.1	-
		36	9.8	12.7	8.2	-
		42	10.4	11.0	7.3	6.4
		50	10.3	9.9	7.2	-
		100	10.7	6.8	5.7	-
		250	10.8	5.9	5.5	-
		500	-	-	-	-
Imhof's approx.			12.1 -	-	-	
	3	30	11.5	24.5	13.7	-
		36	11.9	18.9	11.3	-
		42	12.7	16.6	10.4	5.8
		50	12.3	14.3	9.6	-
		100	12.9	8.4	6.5	-
		250	13.6	6.3	5.8	-
		500	-	-	-	-
Imhof's approx.			14.7 -	-	-	
	4	30	13.5	41.8	23.4	-
		36	14.4	30.8	16.6	-
		42	15.4	24.3	13.8	5.9
		50	15.5	18.9	11.3	-
		100	16.2	9.9	7.4	-
		250	16.7	6.7	5.7	-
		500	-	-	-	-
Imhof's approx.			17.9	-	-	
DGP3	1	100	3.9	7.0	6.3	-
		130	4.2	6.4	6.0	-
		156	3.6	6.6	6.3	4.5
Imhof's approx.			3.6	-	-	
	2	100	26.3	12.5	11.3	-
		130	26.2	10.3	9.3	-
		156	25.3	9.7	8.8	6.4
Imhof's approx.			24.4	-	-	
	3	100	45.6	21.3	18.8	-
		130	45.3	16.1	14.2	-
		156	45.1	13.7	11.9	5.4
			42.1	-	-	

[1] The approximate bias of the LRU statistic is based on Imhof's formula (see section 5.4, p.65 and p.67).

Table 6.4b: Estimated rejection frequencies in % for given $\alpha = 5\%$

DGP	n	T	W_{QS}	W_{PW}	W_{RPW}	W_{VH}
DGP1*	1	42	11.7	11.3	10.3	10.4
		50	11.4	10.6	10.1	10.1
		100	8.9	8.1	7.4	7.8
		250	7.2	6.4	5.5	5.7
		500	6.8	6.1	5.4	5.4
	2	42	17.8	17.6	15.1	13.9
		50	16.2	15.5	13.9	13.3
		100	11.9	10.7	10.1	10.8
		250	9.4	7.0	6.6	6.4
		500	7.3	5.5	4.7	4.2
	3	42		26.8	27.9	22.5	19.5
		50		24.5	25.2	20.6	19.4
		100		16.9	15.9	13.8	13.6
		250		11.1	9.9	9.0	10.0
		500		8.4	7.1	6.9	7.2
	4	42	36.6	37.8	28.2	23.7
		50	33.4	34.0	26.1	22.5
		100	21.8	20.5	16.3	16.1
		250	13.0	11.5	10.2	11.2
		500	9.6	8.3	7.0	6.8
DGP3	1	156	1.1	5.4	5.3	
	2	156	15.0	14.6	14.3	
	3	156	42.2	31.4	31.3	

Note: in the $n=3$ rows, the values 22.5/20.6/13.8/9.0/6.9 fall under W_{RPW} and 19.5/19.4/13.6/10.0/7.2 fall under W_{VH}, while 26.8/24.5/16.9/11.1/8.4 under W_{PW} and 27.9/25.2/15.9/9.9/7.1 under W_{RPW}.

6.4.2 Bias of the LRU test under dynamic misspecification

The simulation experiments on the LRU test are motivated by two questions. First, we are interested in studying the importance and direction of the bias of the LRU test under dynamic misspecification for specific examples. Second, it will be checked how close the small sample distribution of the LRU statistic is to its approximate distribution based on Imhof's formula (see section 5.4, p.67, equation (5.35)). The simulation results for the LRU test are summarised in figures 6.1-6.3a.

Figure 6.1 displays the estimated bias of the LRU test for DGP1* (errors and regressors are VAR(1)) for $n = 1, 2, 3, 4$ and $T = 42$. One can see that the bias is substantial and increases with the number of equations n. For example for $n = 1$ the bias of the LRU test is around 5% for given $\alpha = 5\%$, and for $n = 4$ around 9%. Varying T has only little effect on the bias. The results for other T are, therefore, not reported. The results for DGP1* were, furthermore, compared with those for DGP1 (fixed regressors). The results are very similar and, therefore, the results for DGP1 are not reported.

Figure 6.1a displays the approximate bias of the LRU test for DGP1* for $n = 1, 2, 3, 4$ and $T = 42$, where the approximation is based on Imhof's formula. By comparing the figures 6.1 and 6.1a, one can see that the asymptotic approximation is rather close to the estimated bias by simulation.

Figure 6.2 displays the estimated and figure 6.2a the approximate bias based on Imhof's formula of the LRU test for DGP2* (errors are MA(1) and regressors are VAR(1)) for $n = 1, 2, 3, 4$ and $T = 42$. The estimated bias of the LRU associated with DGP2* are somewhat smaller than those with DGP1*. The bias is positive but its relation to n is not clear. The bias for $n = 1$ is larger than for $n = 2$ and the largest for $n = 4$.

The approximation based on Imhof's formula works well (see figure 6.2a). The results for DPG2 (fixed regressors) are close to those for DGP2* and not reported.

Figure 6.3 plots the estimated and figure 6.3a the approximate bias of the LRU test against α for DGP3 (errors are VAR(4) and regressors are fixed) for $n = 1, 2, 3$ and $T = 156$. The bias increases with n. Surprisingly, the bias is close to zero for $n = 1$ and for $n = 2$ or $n = 3$, the bias is extremely large. The results are similar for other choices of T.

These examples illustrate that testing for homogeneity, where it is falsely assumed that the errors are independent in time, makes inference invalid. In most examples, the null hypothesis of homogeneity is rejected too often and the bias is large. It is a plausible reason why homogeneity has been rejected in many empirical applications. The rejection frequency depends on the number of equations used. This shows that the small sample performance of the dynamically misspecified omitted variable test differs for the univariate and multivariate regression models.

Furthermore, it was illustrated that the asymptotic approximation of the small sample distribution of the LRU statistic under dynamic misspecification (based on Imhof's formula) works well.

**Bias of the LRU test for DGP1* and DGP2*: estimation by simulation
and Imhof's approximation for $T = 42$, $n = 1, 2, 3, 4$**

Figure 6.1: DGP1*

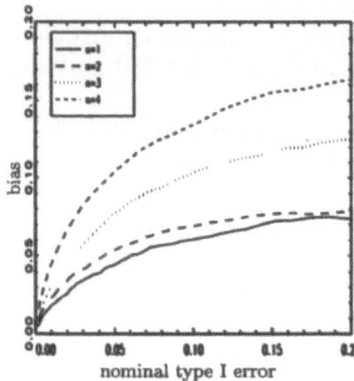

Figure 6.1a: DGP1* (Imhof's approx.)

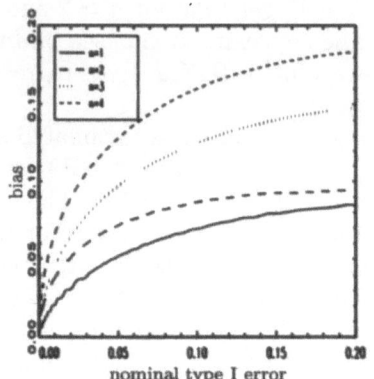

Figure 6.2: DGP2*

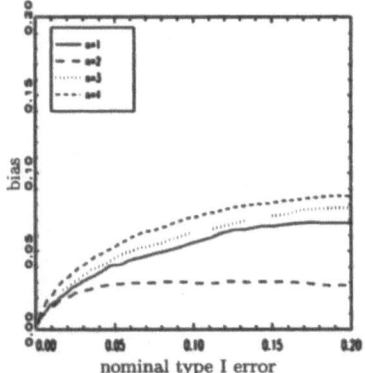

Figure 6.2a: DGP2* (Imhof's approx.)

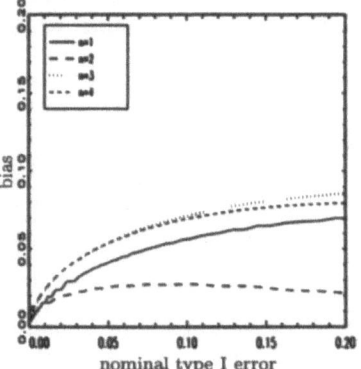

Bias of the LRU test for DGP3: estimation by simulation and Imhof's approximation for $T = 156,\ n = 1, 2, 3$

Figure 6.3: DGP3

Figure 6.3a: DGP3 (Imhof's approx.)

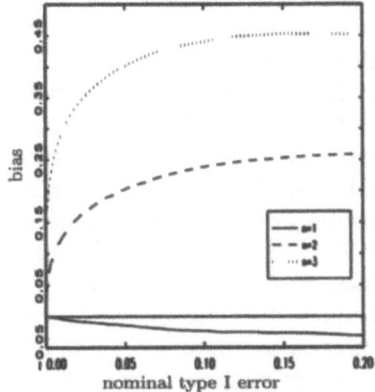

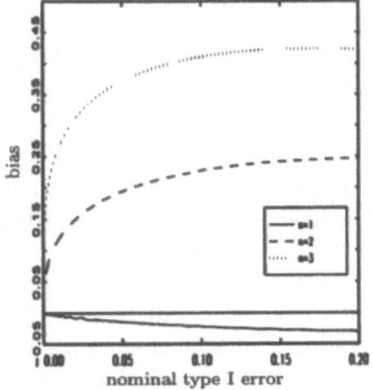

6.4.3 The small sample bias of the correctly specified LR test

The motivation of the experiments here is to study the small sample performance of the dynamically correctly specified LR test for homogeneity. We consider the usual LR test, the small sample corrected version of the LR test (LRC) and the related Monte Carlo test (LR-MC). By varying the sample size, one can see how fast the small sample distribution of the LR and LRC statistic converge to the asymptotic distribution. It will be seen that the latter depends on the number of equations n. The simulation results for the LR and LRC test are summarised in figures 6.4-6.9, for the LR-MC test in figures 6.10-6.11.

Figures 6.4-6.7 show the biases of the LR and LRC test for DGP1* (errors and regressors are VAR(1)) for different n and T. The results do not differ much from those for DGP1 where the regressors are fixed. The estimated bias of the LR test for $n = 1, 2, 3, 4$ and given $T = 42$ is shown in figure 6.4 and for the LRC test in figure 6.5. The number of $T = 42$ corresponds to the realistic case of disposing 42 yearly observations. The bias increases with the number of equations n for both tests. By comparing figures 6.4 and 6.5, one can see that the bias of the LRC test is less pronounced than that of the LR test, suggesting that the small sample version of the LR test is valid. Nevertheless, the bias of the LRC test is substantial for $n > 2$. In figures 6.6 and 6.7, the estimated biases of the LR and LRC test are plotted for given $\alpha = 0.05$ and $n = 1, 2, 3, 4$ against T. As can be seen from these figures, the estimated bias converges slower to zero if n increases.

Figures 6.8-6.9 show the bias of the LR test for DGP3 (errors are VAR(4) and regressors are fixed) for different n and T. The results for the LRC test are similar because T is rather large implying that the small sample correction is very small. These results are not reported, therefore. The estimated bias of the LR test for $n = 1, 2, 3, 4$ and given $T = 100$ is shown in figure 6.8 and for $T = 156$ in figure 6.9. The bias is positive and increases with n in both figures. For $n = 1$ the bias is rather small and for $n = 3$ quite large. By comparing figures 6.8 and 6.9, one can see that the effect of increasing the number of observations on the test's bias is strongly related to the number of equations. Consider for example the change of the bias for $n = 1$ compared with $n = 3$ when moving from $T = 100$ to $T = 156$.

These results indicate that inference based on the LR or LRC test is already problematic for rather small n, where the problem consists in rejecting the null hypothesis of homogeneity too often. This is rather disappointing from a practical point of view. It shows that even if one happens to identify the dynamic structure of the error process correctly, the LR and LRC test for homogeneity in a demand system of moderate size are biased towards rejection. Nevertheless, if one does not reject the null hypothesis of homogeneity, then the simulation results indicate that this conclusion is not problematic. It should be noted, however, that the simulation results are DGP-specific. The observed overrejection of the null hypothesis for the examples here cannot be generalised.

The results for the Monte Carlo test (LR-MC) are summarised in figures 6.10-6.11. Figure 6.10 displays the bias of the LR-MC test for DGP1* (errors and regressors are VAR(1)) for different n and $T = 42$. Figure 6.11 shows the bias of the LR-MC test for DGP3 (errors are VAR(4) and regressors are fixed) for different n and $T = 156$. The simulation results show the excellent performance of the LR-MC test. The estimated true type I error is very close to the nominal type I error.

These results suggest that the omitted variable tests LR and LRC for the kind of model considered here should be interpreted with caution in applied work. The Monte Carlo test should be implemented preferably, as it clearly outperforms the LR and LRC test. The bias of the LR-MC test is small. One disadvantage of the Monte Carlo test is that it is computationally quite costly. The purpose is, however, to do reasonable statistical inference. Since the LR and LRC test do not perform satisfactorily in small samples for $n > 1$, one must choose an alternative testing procedure. A more serious argument against the LR-MC test used here is that its small sample distribution depends on nuisance parameters. The LR-MC test works well with the defined DGP's here but this need not be the case for other DGP's. It should also be noted that the analysis here is restricted to the type I error. The type II error of the LR-MC test is not studied and is an interesting topic for future research.

Estimated bias of the LR and LRC test for DGP1* and $n = 1, 2, 3, 4$

Figure 6.6: LR test, $\alpha = 0.05$

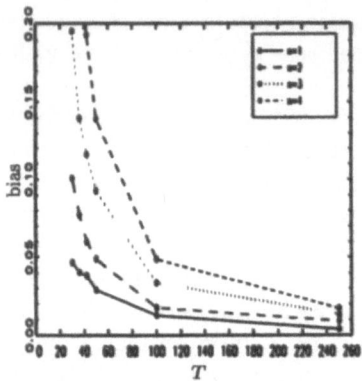

Figure 6.7: LRC test, $\alpha = 0.05$

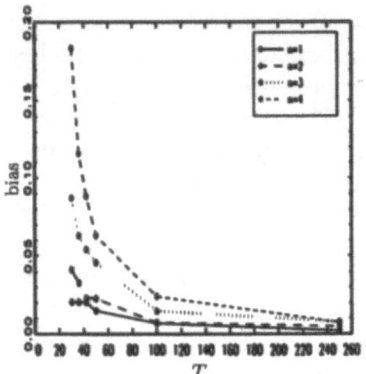

Figure 6.4: LR test, $T = 42$

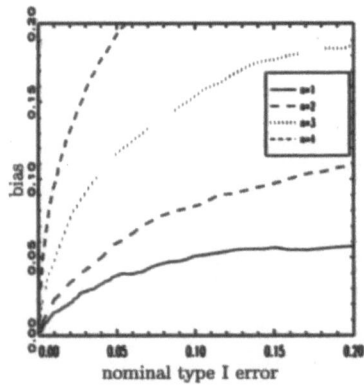

Figure 6.5: LRC test, $T = 42$

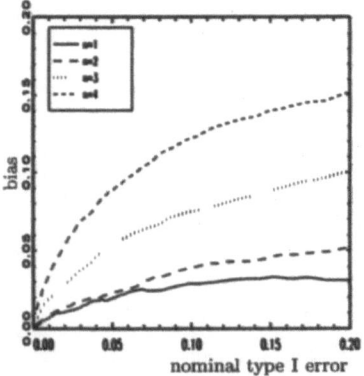

Estimated bias of the LR test for DGP3 and given T

Figure 6.8: LR test, $T = 100$

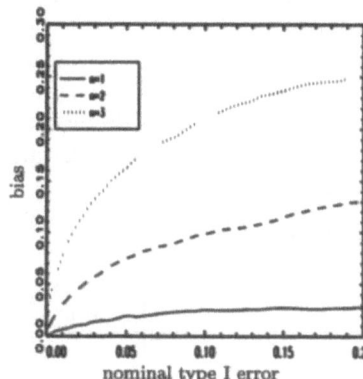

Figure 6.9: LR test, $T = 156$

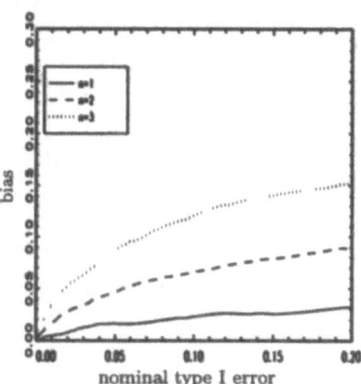

Estimated bias of the LR-MC test for given DGP and T

Figure 6.10 DGP1*, $T = 42$

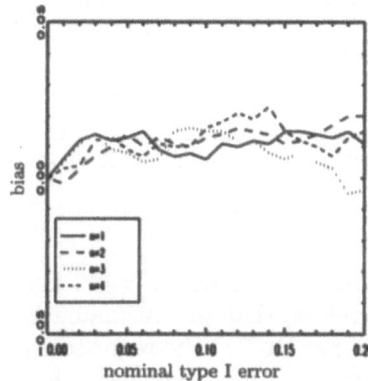

Figure 6.11: DGP3, $T = 156$

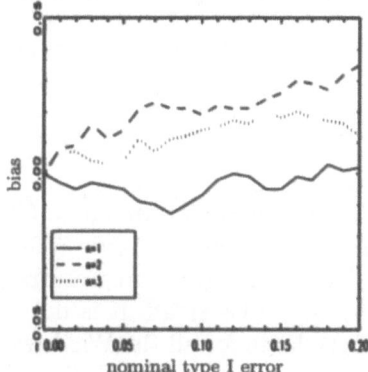

6.4.4 The small sample bias of the robust Wald test

The motivation of the simulation experiments here is to evaluate the bias of the various robust Wald tests for specific examples. The Wald test was denoted in general by W_{HAC} and is defined in section 5.5, p.71. The various versions of the Wald tests depend on the specific robust variance-covariance matrix estimator and are denoted by $W_{QS}, W_{PW}, W_{RPW}, W_{VH}$ tests. In order to take the small sample problem into account, we have defined the F-versions of these tests. The different Wald tests are summarised in table 6.5 and the simulation results are summarised in figures 6.12-6.27.

Table 6.5: Definition of the robust Wald tests

Statistic	based on the HAC estimator	critical value
W_{QS} WF_{QS}	[1] quadratic spectral (QS), section 4.2.2, p.40 F-version of W_{QS}, see eq. (5.43), p.72	χ_α F_α
W_{PW} WF_{PW}	[2] prewhitened quadratic spectral (PW-QS), section 4.2.2, p.41 F-version of W_{PW}, see eq. (5.43), p.72	χ_α F_α
W_{RPW} WF_{RPW}	restricted PW-QS, section 6.3.4, p.90 F-version of W_{RPW}, see eq. (5.43), p.72	χ_α F_α
W_{VH} WF_{VH}	[3] VARHAC (VH), section 4.2.2, p.42 F-version of W_{VH}, see eq. (5.43), p.72	χ_α F_α

χ_α satisfies $P(\chi_n^2 \leq \chi_\alpha) = 1-\alpha$ and F_α satisfies $P(F_{n;T-2n-2} \leq F_\alpha) = 1-\alpha$.
[1] see Andrews (1991),
[2] see Andrews and Monahan (1992),
[3] see Den Haan and Levin (1997).

For DGP1 (fixed regressors), the number of observations $T = 42$ is admittedly quite small. It is not surprising, therefore, that the estimated bias is very large for all the Wald tests in this case. For DGP1* the regressors are generated randomly. For $T = 42$, the estimated rejection frequencies based on DGP1* and DGP1 are rather close and the results for DGP1 are not reported, therefore.

For DGP1*, the estimated bias is plotted against α for the W_{QS}, W_{PW}, W_{RPW}, W_{VH} tests for $T = 42$ and $n = 1, 2, 3, 4$ in figures 6.12-6.15. The large biases indicate the poor performance of these tests. This is even true for $n = 1$. The results indicate that the bias of the W_{HAC} test increases substantially with the number of equations. The related F-versions of the Wald tests $WF_{QS}, WF_{PW}, WF_{RPW}$ and WF_{VH} perform somewhat better than their uncorrected counterparts. The bias corrections induced by the

F-version of the Wald tests is very small, nevertheless, compared with the overall bias and these results are not reported, therefore.

Figures 6.16-6.19 display the estimated biases of the W_{QS}, W_{PW}, W_{RPW} and W_{VH} test for different T and given n and $\alpha = 0.05$. Here the number of observations T is varied between 42 and 500 in order to see for which sample size the robust Wald test is reliable. One can see from these figures that the bias goes quite slowly to zero. For $n = 4$, the Wald test is seriously biased even for $T = 250$. For $n = 4$ and $T = 500$ the biases of the W_{RPW} and W_{VH} are reasonably small; this is not the case for the W_{QS} and W_{PW} test. This result is disillusioning. It suggests that the robust Wald test should not be applied in multivariate regression problems unless the number of observations is extremely large. Note that DGP1* has only simple dynamic features: regressors (except the constant) and errors are generated by two independent stationary VAR(1) processes.

Next, the performance of the Wald tests is investigated for DGP3. Recall that DPG3 is motivated as a "realistic" DGP for the seasonal Rotterdam model. For empirical work, one might think that 156 seasonal observations are sufficient for reasonable inference based on the robust Wald test. After having discussed the results for DGP1*, it is of course not surprising that this is not the case. Figures 6.20-6.23 display the estimated biases of the $W_{QS}, W_{PW}, W_{RPW}, W_{VH}$ tests for $n = 1, 2, 3$ and $T = 156$. The bias increases with the number of equations n. For $n = 1$, the bias is negative for the W_{QS} and W_{PW} test. This illustrates that the bias of these tests is not always positive as the other simulation results might have suggested.

It should be noted that the W_{RPW} test with the restricted VAR(1) prewhitening procedure performs better than the W_{PW} test with the unrestricted VAR(1) prewhitening procedure in the examples above. This is especially true for larger n. This illustrates that the prewhitening procedure can have an impact on the small sample behaviour of the test statistic.

Interestingly, the W_{VH} and W_{RPW} test perform better than the other two robust Wald tests analysed in this section. Since the DGP's are quite specific, this result cannot be generalised.

The simulation results for DGP1* and DGP3 suggest that the number of observations is not sufficient to do correct inference with the robust Wald test when the number of equations n exceeds one. For the univariate regression model, the simulation results are mixed. The "robust" Wald statistic is approximately distributed as a chi-square only for rather large T even if $n = 1$.

There are several reasons why the "robust" approach for testing homogeneity does not work well in this context. It is important to note that the QS and PW-QS estimator are constructed to minimise the asymptotic mean squared error of an estimated asymptotic variance. The relation between this optimality criterion and the test statistic is not clear (see Andrews 1991, p.828). Nevertheless, it is reasonable to assume that a poor estimate of the asymptotic variance causes a poor performance of the W_{HAC} test. Therefore, some possible factors, which may cause the poor estimation of the asymptotic variance, will be discussed below.

For the QS and PW-QS estimator, one problem may be related to the estimation of the bandwidth parameter. In contrast to the simulation experiments of Andrews and Monahan (1992), *one* bandwidth parameter has to be used for estimating the variances of *several* estimated parameters. This problem is discussed in Den Haan and Levin (1997). The number of estimated variances needed for the Wald test can be reduced if estimation is based on the transformed model equation $y_t = B^* x_t^* + u_t$ (see equation (5.3)). As shown in section 5.1, the null hypothesis of homogeneity implies that the last column of the coefficient matrix B^* is a zero vector. Den Haan and Levin (1997) propose to give weight only to the parameters of interest. This approach did not improve the performance of the semi-parametric tests and, therefore, these results are not reported.

One result from the experiments from Andrews and Monahan(1992) and Den Haan and Levin (1997) was that the HAC estimators and the W_{HAC} tests perform worse if the temporal dependence of the error process increases. It will, therefore, be checked if the specific dynamic structure of the error process could explain (at least to a large extent) why the W_{HAC} test performs badly in our example. The experiments for DGP1* were repeated for a modified DGP1*. The error process for DGP1* was defined by $u_t = R u_{t-1} + \epsilon_t$ with $\epsilon_t \sim i.i.d.N(O, \Sigma)$. Now, the errors are generated by a white noise process with $u_t = \epsilon_t \sim i.i.d.N(O, \Sigma)$. The modified DGP is identical to DGP1* if R is replaced by the zero matrix. Recall that the regressors (without the constant terms) are generated by a VAR(1) process.

In figures 6.24-6.27, the results for the $W_{QS}, W_{PW}, W_{RPW}, W_{VH}$ tests for the modified version of DGP1* (errors are white noise) are shown. In these figures, the estimated bias of the W_{HAC} test is plotted against T for given $\alpha = 0.05$ and $n = 1, 2, 3, 4$. By comparing the figures 6.24-6.27 with 6.19-6.22, one can see that the test's biases are smaller if the errors are spherical compared to the VAR(1) error process. It is interesting to note that for increasing n, the estimated biases are decreasing faster in T if the errors are white noise compared to the VAR(1) error process. Nevertheless, the bias is still quite large even if the errors are white noise and if $T = 250$ and $n > 2$. This is also true for $n = 1$ and $T = 100$. The specific dynamic structure can, therefore, explain only partly the bad performance of the W_{HAC} test. The "robust" Wald test performs badly even if the errors are spherical. These results strongly suggest that the "robust" Wald test should not be applied to a multivariate time-series model. Furthermore, the simulation results indicate that the small sample inference for univariate regression models based on the "robust" Wald test should interpreted with caution.

Estimated bias of the W_{HAC} test for DGP1* and $T = 42$, $n = 1, 2, 3, 4$

Figure 6.12: W_{QS} test

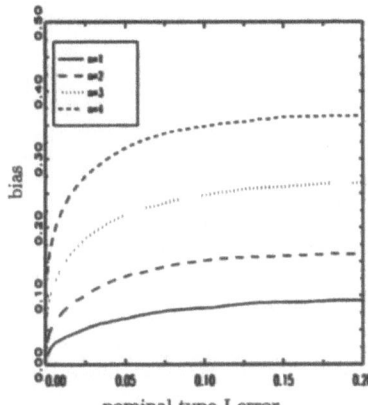

Figure 6.13: W_{PW} test

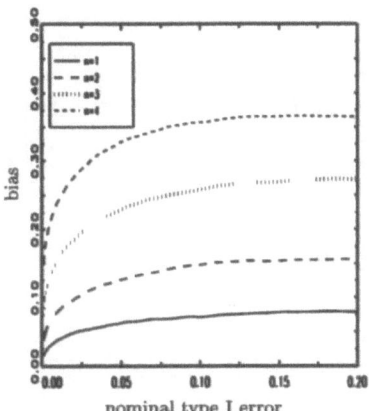

Figure 6.14: W_{RPW} test

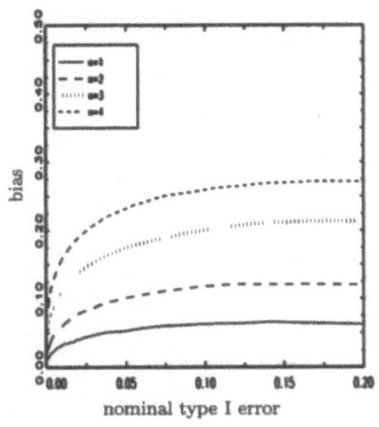

Figure 6.15: W_{VH} test

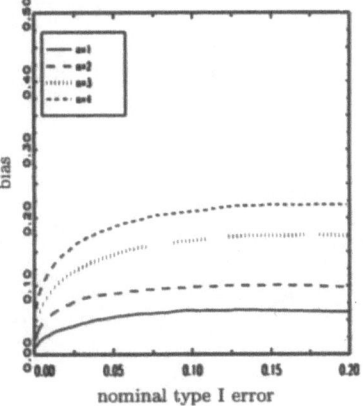

Estimated bias of the W_{HAC} test for DGP1*
and nominal type I error of 5%

Figure 6.16: W_{QS} test

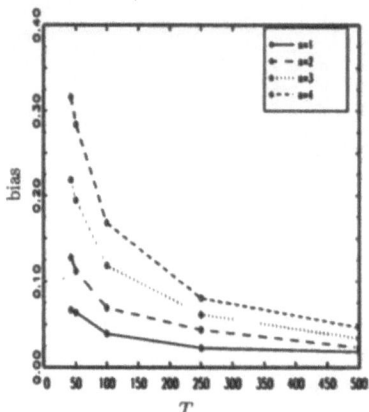

Figure 6.17: W_{PW} test

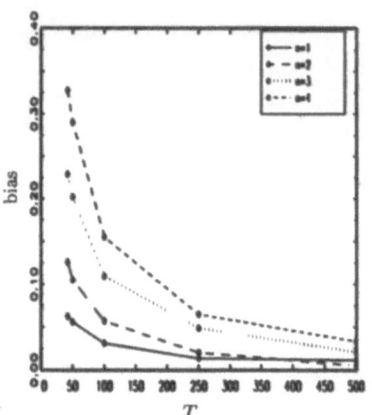

Figure 6.18: W_{RPW}

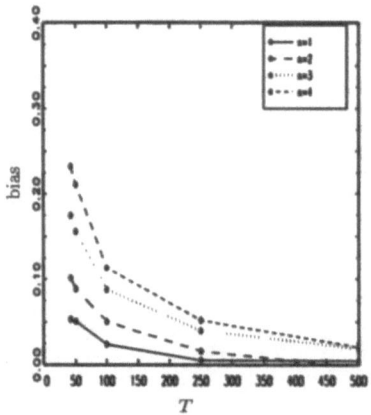

Figure 6.19: W_{VH} test

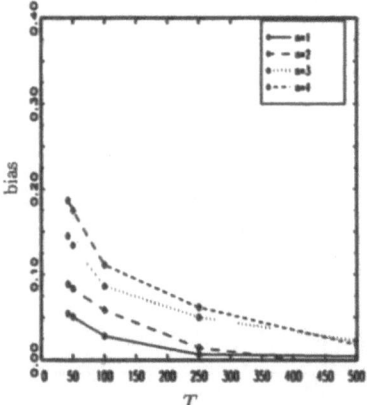

Estimated bias of the W_{HAC} test for DGP3 and $T = 156$, $n = 1, 2, 3$

Figure 6.20: W_{QS} test

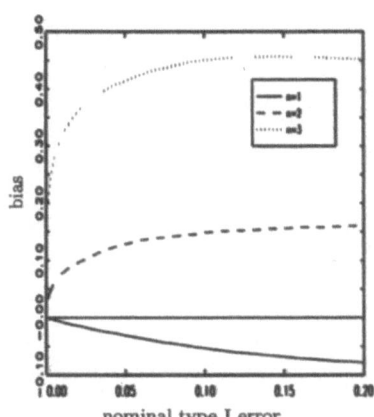

Figure 6.21: W_{PW} test

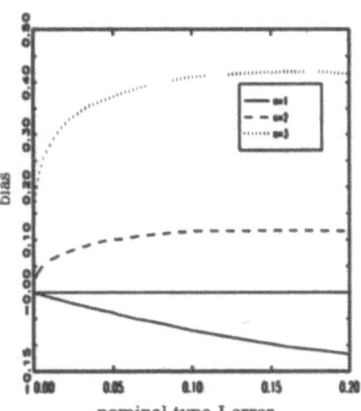

Figure 6.22: W_{RPW} test

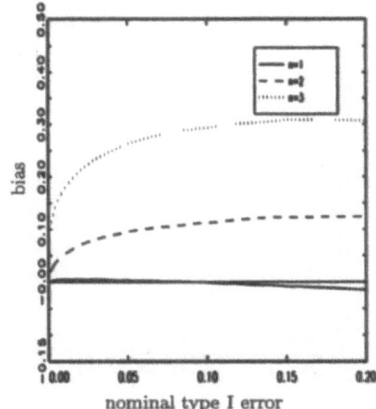

Figure 6.23: W_{VH} test

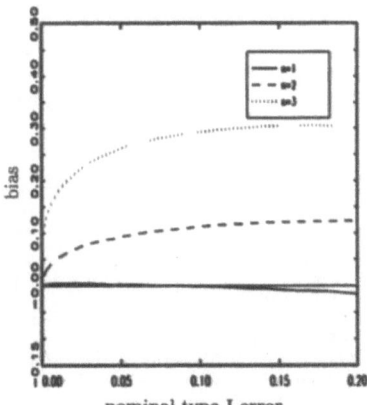

Estimated bias of the W_{HAC} test with *spherical* errors and nominal type I error of 5%

Figure 6.24: W_{QS} test

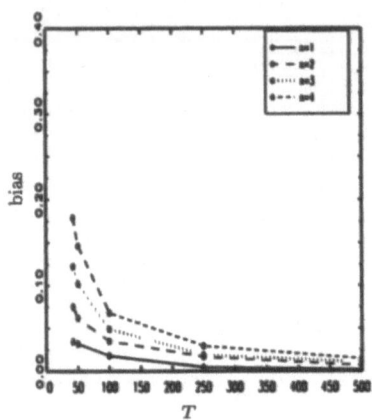

Figure 6.25: W_{PW} test

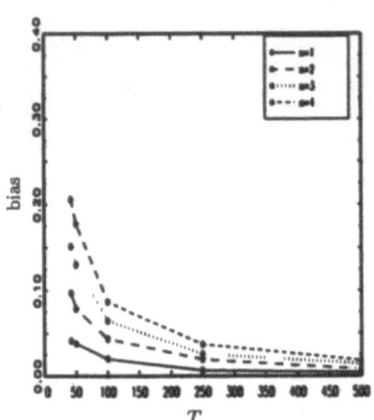

Figure 6.26: W_{RPW}

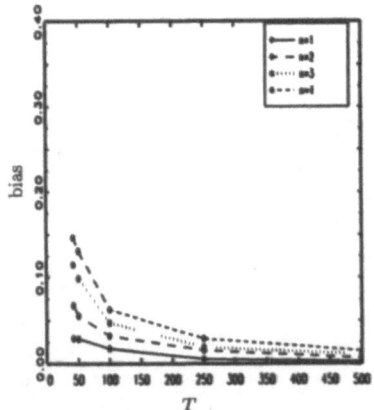

Figure 6.27: W_{VH} test

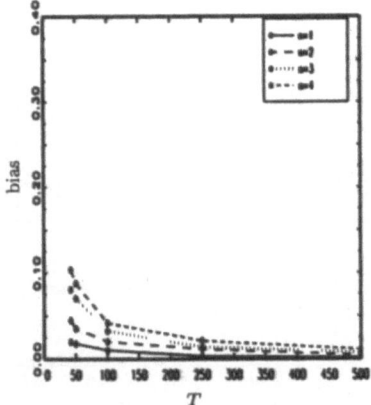

6.4.5 Summary

In this chapter, the performance of the various tests for homogeneity discussed in chapter 5 were analysed by Monte Carlo experimentation. We have studied various versions of the homogeneity test when the test statistics are dynamically misspecified and when they are correctly specified. As mentioned before, the simulation analysis can be seen as an investigation of the omitted variable test for a linear multivariate regression system with dynamic errors and a specific data-generating process. The small sample properties of the various test statistics for homogeneity depend on so many parameters that the analysis was restricted to a small set of DGP's. In order to define some "realistic" DGP's, the population parameters were based on the estimation results for the Rotterdam model with annual and quarterly data from the UK. The simulation results can be summarised as follows:

- Anderson's $\mathcal{U}$ test (LRU) under *dynamic misspecification*: In most of the examples the homogeneity test is strongly biased towards rejection and the bias increases with the number of equations. This illustrates that the results for the univariate and multivariate regression models can differ substantially. As an alternative method of estimating the test's bias by simulation, one can approximate the bias by Imhof's formula. The simulation results illustrate the necessity of using a correctly specified test for homogeneity.
- Likelihood ratio tests (LR, LRC, LR-MC): The small sample performance of the usual LR test is not satisfactorily, especially when the number of equations is large. As a remedy, the small sample performance of two related likelihood ratio tests were investigated: the modified likelihood ratio test (LRC) and the Monte Carlo test (LR-MC). The small sample bias of the LRC test is smaller than that of the LR test. The bias reduction is, however, too small for useful inference. The small sample bias of the Monte Carlo test is small in all examples. Although the LR-MC test is computationally quite costly to implement, it is strongly recommended.
- Robust Wald tests: The robust Wald test is based on the quasi-maximum likelihood estimator and a HAC estimator of the variance-covariance matrix. We have studied the robust Wald tests based on the (prewhitened) quadratic spectral estimator proposed in Andrews (1991) and Andrews and Monahan (1992) and the VARHAC estimator proposed in Den Haan and Levin (1997). The small sample biases of these tests are very large in most of our examples. The bias increases with the number of equations and converges very slowly to zero when the number of observations increases. The performance of the related F-versions of the robust Wald tests is only a little bit better. The conclusion from the simulation study is that if the number of equations exceeds one, the robust Wald test should not be applied at all unless the number of observations is extremely large. For annual and seasonal demand systems the sample size is too small for reliable inference to be based on the robust Wald test.

7. Conclusions

In this work, we have studied the omitted variable test for a dynamic multivariate regression model. The analysis was applied to the homogeneity test in a demand system (the Rotterdam model). In order to emphasise the importance of taking the dynamics of the regression model into account, we have studied the omitted variable test under *dynamic misspecification* first.

As a simple example, the performance of the quasi t-test was analysed in detail for a univariate regression model with one AR(1) regressor and an AR(1) error term. It was shown that the distribution of the t-statistic depends on the autoregressive parameters of the processes for the regressor and the error term and on the sample size.

For the multivariate regression model the omitted variable test under dynamic misspecification has not been analysed previously. It was shown that the distribution of the F-version of the Wald test is asymptotically distributed as a quadratic form in normal variables and its distribution can be calculated by Imhof's formula. This general result can be applied to various dynamic models in order to approximate the test's bias under dynamic misspecification.

We applied the analysis to the homogeneity test in the Rotterdam model. For the simulation study various data-generating processes were defined which are based on the Rotterdam model estimated with annual and seasonal data from the UK. The results in the simulation study suggest that the bias of the homogeneity test increases with the number of equations in most of the cases. Here the number of equations varied between one and four. It was found, furthermore, that the approximation based on Imhof's formula works very well.

The conclusion from these results is that the correct dynamic specification of the regression model is very important for correct statistical inference. The empirical researcher should check carefully if his model is not dynamically misspecified. This is, of course, a difficult task and was not discussed here (see, e.g., Deschamps, 1993).

Having illustrated the importance of using a correctly specified omitted variable test, two classes of omitted variable tests are considered. These are the robust Wald tests based on the heteroskedasticity and autocorrelation consistent variance-covariance matrix estimators and versions of the likelihood ratio test.

The motivation for using the *robust Wald* statistics is that they have an asymptotic chi-square distributions under quite general conditions. The researcher does, therefore, not need to specify the dynamics of the errors process. The cost of this approach is, however, very large. The most important conclusion from the simulation study is that the small sample performance of the robust Wald tests is so bad that they should not be used at all for testing homogeneity. It was illustrated that the number of observations needed for reliable inference is usually extremely large. For example, even if the errors are white noise, the hypothesis of homogeneity was rejected approximately twice too often for a regression system with $n = 4$ and a sample size of 250 for given nominal size of 5%. Taking into account that the sample size with annual and seasonal data does not usually exceed 250, the message is clear: do not use these robust Wald tests in this context. The simulation results for the univariate regression model $(n = 1)$ are mixed. In some cases the small sample bias of the robust Wald test is very small but in other cases it is quite large. The simulation results indicate, furthermore, that the temporal dependence in the errors increases the bias of the robust Wald test.

As an alternative to the robust Wald tests, we have investigated the performance of the *likelihood ratio* test. Since only the asymptotic distribution of the likelihood ratio statistic is known, the small sample bias of this test was analysed by simulation. The simulation results indicate that the homogeneity test is biased towards rejection. As before, the test's bias increases with number of equations. This result shows again that the small sample bias of the omitted variable test differs for univariate and multivariate regression models. In all the examples considered here, the small sample problem increases with the number of equations.

In order to tackle the small sample problem two remedies were proposed: a small sample corrected likelihood ratio test and a Monte Carlo test. The small sample corrected likelihood ratio test performs better than its uncorrected counterpart but is still unsatisfactorily. The Monte Carlo test proposed here can be interpreted as a parametric bootstrap version of the likelihood ratio test. Its small sample performance is excellent for all the examples studied and it is, therefore, recommended.

The general conclusions described above lead us to the following practical recommendations for the researcher using the omitted variable test when estimating a multivariate regression model:

- Check the dynamic specification of the model carefully.
- Do *not* use the robust Wald test based on the heteroskedasticity and autocorrelation consistent (HAC) variance-covariance matrix estimator.
- Use the parametric bootstrap version of the likelihood ratio test.

A. Proof of proposition 5.4

It was assumed that the errors are covariance-stationary with absolutely summable autocovariances and that $M_j := \lim_{T\to\infty} E(\sum_{t=j+1} x_t x'_{t-j}/T)$ with the M_j's being absolutely summable. Furthermore, x_t is uncorrelated with u_τ for any t, τ.

We are looking for the eigenvalues of the matrix F:

$$e_i(F) = e_i\left((a'M_0^{-1}a)^{-1}\Sigma_u R_H \Phi R'_H\right) \quad,$$

where

$$\Phi := (M_0^{-1} \otimes I_n)V(M_0^{-1} \otimes I_n) \quad,$$

$$V := \lim_{T\to\infty} Var(T^{-1/2}\sum_{t=1}^{T} v_t)$$

and

$$R_H := a' \otimes I_n \quad.$$

Calculation of V: Recall that $v_t := x_t \otimes u_t$ and note that $E(v_t) = 0$. The asymptotic variance V can be expressed as

$$
\begin{aligned}
V \; &:= \; \lim_{T\to\infty} Var(T^{-1/2}\sum_{t=1}^{T} v_t) = \lim_{T\to\infty}\frac{1}{T}E\left(\sum_{t=1}^{T} v_t\right)E\left(\sum_{t=1}^{T} v'_t\right) \\
&= \; \lim_{T\to\infty}\frac{1}{T}E\left(\sum_{t,s=1}^{T} v_t v'_s\right) \\
&= \; \lim_{T\to\infty}\frac{1}{T}[\sum_{t=1}^{T} E(v_t v'_t) + \sum_{t=2}^{T}\left(E(v_t v'_{t-1}) + E(v_{t-1}v'_t)\right) + \cdots + \\
&\quad (E(v_T v'_1) + E(v_1 v'_T))] \\
&= \; \lim_{T\to\infty}\frac{1}{T}\left[\sum_{t=1}^{T} E(v_t v'_t) + \sum_{j=1}^{T-1}\sum_{t=j+1}^{T}\left(E(v_t v'_{t-j}) + E(v_{t-j}v'_t)\right)\right] \quad.
\end{aligned}
$$

Note that

$$
\begin{aligned}
E(v_t v'_{t-j}) \; &= \; E\left((x_t \otimes u_t)(x'_{t-j} \otimes u'_{t-j})\right) = E\left((x_t x'_{t-j}) \otimes (u_t u'_{t-j})\right) \\
&= \; E(x_t x'_{t-j}) \otimes E(u_t u'_{t-j}) \quad,
\end{aligned}
$$

where the last line follows from the law of iterated expectations.
By assumption, $M_j := \lim_{T\to\infty} E(\sum_{t=j+1}^{T} x_t x'_{t-j}/T)$ and let
$\Sigma_{u,j} := E(u_t u'_{t-j})$.
Therefore,

$$\lim_{T\to\infty} \frac{1}{T} E(\sum_{t=j+1}^{T} v_t v'_{t-j}) = \left(\lim_{T\to\infty} E(\sum_{t=j+1}^{T} \frac{x_t x'_{t-j}}{T}) \right) \otimes \Sigma_{u,j} = M_j \otimes \Sigma_{u,j} \quad .$$

Then, V can be expressed as

$$V = (M_0 \otimes \Sigma_{u,0}) + \sum_{j=1}^{\infty} \left((M_j \otimes \Sigma_{u,j}) + (M'_j \otimes \Sigma'_{u,j}) \right) \quad .$$

Calculation of Φ: Recall that $\Phi := (M_0^{-1} \otimes I_n)V(M_0^{-1} \otimes I_n)$. Replacing
V found above yields

$$
\begin{aligned}
\Phi \quad &:= \quad (M_0^{-1} \otimes I_n)V(M_0^{-1} \otimes I_n) \\
&= \quad (M_0^{-1} \otimes I_n) \left((M_0 \otimes \Sigma_{u,0}) + \sum_{j=1}^{\infty} \left((M_j \otimes \Sigma_{u,j}) + (M'_j \otimes \Sigma'_{u,j}) \right) \right) \\
&\qquad (M_0^{-1} \otimes I_n) \\
&= \quad M_0^{-1} \otimes \Sigma_{u,0} + \sum_{j=1}^{\infty} \left((M_0^{-1} M_j M_0^{-1} \otimes \Sigma_{u,j}) + (M_0^{-1} M'_j M_0^{-1} \otimes \Sigma'_{u,j}) \right)
\end{aligned}
$$

Calculation of $R_H \Phi R'_H$: Recall that $R_H := a' \otimes I_n$, where a is a column
vector. Substitution of Φ found above yields

$$
\begin{aligned}
R_H \Phi R'_H &= (a' \otimes I_n)\Phi(a \otimes I_n) \\
&= \left(a' M_0^{-1} a \otimes \Sigma_{u,0} \right) + \\
&\quad \sum_{j=1}^{\infty} \left((a' M_0^{-1} M_j M_0^{-1} a \otimes \Sigma_{u,j}) + (a' M_0^{-1} M'_j M_0^{-1} a \otimes \Sigma_{u,j}) \right) \\
&= \left(a' M_0^{-1} a \Sigma_{u,0} \right) + \sum_{j=1}^{\infty} \left((a' M_0^{-1} M_j M_0^{-1} a \Sigma_{u,j}) + (a' M_0^{-1} M'_j M_0^{-1} a \Sigma'_{u,j}) \right)
\end{aligned}
$$

Using

$$a' M_0^{-1} M_j M_0^{-1} a = a' M_0^{-1} M'_j M_0^{-1} a$$

which follows from the fact that a scalar is equal to its transpose, $R_H \Phi R'_H$
can be written as

$$R_H \Phi R'_H = (a' M_0^{-1} a \Sigma_{u,0}) + \sum_{j=1}^{\infty} \left((a' M_0^{-1} M_j M_0^{-1} a)(\Sigma_{u,j} + \Sigma'_{u,j}) \right) \quad .$$

Calculation of $e_i(F) = e_i\left((a'M_0^{-1}a\Sigma_{u,0})^{-1}R_H\Phi R_H'\right)$:

$$
\begin{aligned}
e_i(F) &= e_i\left((a'M_0^{-1}a\Sigma_{u,0})^{-1}R_H\Phi R_H'\right) \\
&= e_i\left(I_n + \sum_{j=1}^{\infty}\left(\frac{a'M_0^{-1}M_jM_0^{-1}a}{a'M_0^{-1}a}\Sigma_{u,0}^{-1}(\Sigma_{u,j}+\Sigma_{u,j}')\right)\right) \quad,
\end{aligned}
$$

which is the intended result.

B. Data

The following tables contain the population data and the nominal and real consumption household expenditure (component categories) used for the estimation of the Rotterdam model. The data are taken from the Central Statistical Office (UK):

- annual population data: Statistical yearbooks 1977, 1991 and 1996
- annual expenditure data: Economics Trends, Annual Supplement 1996
- seasonal expenditure data: Economics Trends, Annual Supplement 1996

Annual data for the United Kingdom from 1952 to 1995: Population and nominal household consumption expenditure in million £

Year	Pop.	nominal expenditure				
		food	alcohol/ tobacco	clothing/ footwear	energy	other goods
1952	50430	2824	1600	1097	485	1032
1953	50593	3122	1632	1115	522	1074
1954	50765	3295	1649	1205	570	1139
1955	50946	3585	1713	1297	634	1243
1956	51184	3787	1801	1378	716	1333
1957	51430	3928	1887	1439	748	1414
1958	51652	4028	1940	1458	843	1479
1959	51956	4157	1981	1525	867	1559
1960	52372	4225	2103	1664	963	1663
1961	52807	4366	2292	1729	1041	1781
1962	53292	4560	2403	1771	1192	1843
1963	53625	4689	2518	1873	1309	1922
1964	53991	4889	2733	1971	1374	2074
1965	54350	5059	2927	2099	1546	2206
1966	54643	5297	3130	2154	1688	2339
1967	54959	5485	3251	2219	1804	2466
1968	55214	5696	3448	2375	2042	2765
1969	55461	6035	3723	2505	2225	2910
1970	55632	6429	4020	2753	2339	3208
1971	55928	7105	4284	2990	2547	3600
1972	56097	7614	4718	3370	2874	4146
1973	56223	8751	5361	3860	3129	4725
1974	56236	10028	6144	4498	3966	5698
1975	56226	12313	7583	5206	5119	6769
1976	56216	14459	8806	5797	6057	7712
1977	56190	16596	10173	6630	6895	8946
1978	56178	18373	11168	7832	7223	10610
1979	56240	20988	12899	9168	8846	12583
1980	56330	23655	14776	9873	11001	14601
1981	56352	24946	16667	10155	13422	15803
1982	56318	26490	17884	10925	15027	17212
1983	56377	28061	19479	12120	16220	18764
1984	56506	29274	20938	13168	16973	20637
1985	56685	30657	22657	14912	18578	23054
1986	56852	32574	23889	16646	18219	25921
1987	57009	34402	25116	17848	18628	28935
1988	57158	36491	26584	19023	19291	32636
1989	57358	39143	27755	19847	20460	36271
1990	57561	41817	30008	20876	22422	39566
1991	57808	44044	32680	21412	24955	41298
1992	58006	45243	33553	22097	25399	43161
1993	58191	46234	34775	23528	26136	45799
1994	58395	47048	36488	24839	26857	47656
1995	58606	48850	38010	25801	27192	48903

Annual data for the United Kingdom from 1952 to 1995: Real household consumption expenditure in million £, 1990 prices

Year	Population	food	alcohol/ tobacco	clothing/ footwear	energy	other goods
		real expenditure, 1990 prices				
1952	50430	27336	17597	5658	8108	8971
1953	50593	28911	17878	5800	8176	9706
1954	50765	29513	18041	6217	8522	10398
1955	50946	30558	18602	6670	8676	11189
1956	51184	31051	18861	6935	9013	11434
1957	51430	31518	19242	7125	8800	11657
1958	51652	31898	19537	7154	9403	12332
1959	51956	32508	20415	7542	9264	13170
1960	52372	33206	21331	8123	10084	14076
1961	52807	33757	22241	8301	10231	14750
1962	53292	34095	22004	8264	11238	15185
1963	53625	34274	22959	8630	12074	15974
1964	53991	34695	23565	8993	12308	16618
1965	54350	34673	22929	9330	13154	17060
1966	54643	35054	23717	9325	13747	17595
1967	54959	35631	24296	9471	14040	18257
1968	55214	35817	25044	9991	14833	18895
1969	55461	35919	25244	10147	15422	18864
1970	55632	36280	26349	10587	15738	19102
1971	55928	36312	26975	10771	15839	19671
1972	56097	36248	28882	11359	16891	21623
1973	56223	37120	31875	11879	17703	23628
1974	56236	36470	32189	11733	17635	24147
1975	56226	36480	31576	11895	17244	23177
1976	56216	36866	31490	11966	17625	23118
1977	56190	36547	31119	12144	17939	23239
1978	56178	37217	32000	13274	18364	24898
1979	56240	38046	32922	14220	18992	25564
1980	56330	38095	31706	14064	18742	24954
1981	56352	37849	30222	13934	18808	25077
1982	56318	37942	28963	14447	18907	25635
1983	56377	38582	29632	15441	18932	26296
1984	56506	37925	29695	16261	19299	27387
1985	56685	38402	29819	17615	20191	28712
1986	56852	39610	29659	19169	21420	31035
1987	57009	40621	29971	20204	21871	33490
1988	57158	41542	30389	20780	22482	36308
1989	57358	42247	30236	20662	22335	38486
1990	57561	41817	30008	20876	22422	39566
1991	57808	41869	29004	20817	23151	38550
1992	58006	42384	27676	21455	22889	38739
1993	58191	42801	27206	22665	23021	40225
1994	58395	43366	27559	23850	22720	41415
1995	58606	43519	27517	24740	22209	41570

Seasonal data for the United Kingdom from 1952 to 1995: household consumption expenditure in million £

Year		nominal exp.				real exp., 1990 prices		
	food	clothing/ footwear	energy	other goods	food	clothing/ footwear	energy	other goods
1955 Q1	851	262	170	271	7395	1355	2688	2468
Q2	896	332	155	285	7785	1717	2108	2660
Q3	878	301	145	315	7538	1551	1706	2927
Q4	960	402	164	372	7840	2047	2174	3134
1956 Q1	909	281	195	293	7610	1419	2721	2586
Q2	967	347	177	310	7863	1747	2209	2710
Q3	914	327	160	334	7501	1649	1763	2953
Q4	997	423	184	396	8077	2120	2320	3185
1957 Q1	929	297	196	319	7571	1482	2655	2560
Q2	977	368	182	324	7911	1830	2119	2769
Q3	978	340	168	352	7713	1678	1706	3019
Q4	1044	434	202	419	8323	2135	2320	3309
1958 Q1	964	302	238	331	7769	1484	2907	2785
Q2	1013	368	210	344	7996	1796	2344	2895
Q3	985	339	181	370	7835	1669	1736	3152
Q4	1066	449	214	434	8298	2205	2416	3500
1959 Q1	1008	309	254	342	7872	1531	3045	2927
Q2	1039	388	208	370	8216	1926	2159	3160
Q3	1015	348	186	385	7974	1718	1698	3301
Q4	1095	480	219	462	8446	2367	2362	3782
1960 Q1	1031	330	269	368	8136	1614	3156	3134
Q2	1065	426	222	393	8387	2097	2261	3368
Q3	1033	388	211	417	8136	1898	1946	3566
Q4	1096	520	261	485	8547	2514	2721	4008
1961 Q1	1056	358	294	392	8269	1733	3229	3417
Q2	1094	428	243	421	8481	2064	2362	3525
Q3	1091	404	229	451	8319	1939	1973	3733
Q4	1125	539	275	517	8688	2565	2667	4075
1962 Q1	1112	350	339	412	8416	1660	3636	3468
Q2	1154	443	292	436	8548	2072	2703	3667
Q3	1114	425	258	466	8260	1967	2084	3883
Q4	1180	553	303	529	8871	2565	2815	4167
1963 Q1	1143	364	395	416	8339	1682	3622	3506
Q2	1183	457	305	474	8587	2109	2853	3937
Q3	1150	453	282	490	8553	2086	2605	4046
Q4	1213	599	327	542	8795	2753	2994	4485
1964 Q1	1190	394	388	452	8630	1808	3465	3688
Q2	1228	483	340	511	8685	2203	3085	4093
Q3	1192	467	298	521	8452	2158	2705	4182
Q4	1279	627	348	590	8928	2824	3053	4655
1965 Q1	1217	427	432	488	8450	1912	3663	3786
Q2	1261	513	382	541	8653	2285	3290	4157
Q3	1246	505	345	556	8552	2241	2968	4275
Q4	1335	654	387	621	9018	2892	3233	4842

1966 Q1	1282	450	468	521	8643	1974	3838	3965
Q2	1334	529	423	581	8745	2298	3488	4352
Q3	1311	511	365	589	8661	2210	2971	4435
Q4	1370	664	432	648	9005	2843	3450	4843
1967 Q1	1327	461	498	537	8673	1971	3886	4004
Q2	1370	530	438	599	8862	2258	3475	4447
Q3	1370	531	387	620	8926	2273	3038	4596
Q4	1418	697	481	710	9170	2969	3641	5210
1968 Q1	1387	490	573	620	8835	2078	4232	4409
Q2	1410	571	481	657	8913	2412	3522	4443
Q3	1416	569	458	699	8893	2390	3324	4733
Q4	1483	745	530	789	9176	3111	3755	5310
1969 Q1	1461	507	644	638	8836	2102	4480	4255
Q2	1504	618	500	701	8935	2515	3500	4582
Q3	1493	597	466	736	8917	2404	3245	4716
Q4	1577	783	615	835	9231	3126	4197	5311
1970 Q1	1525	547	676	688	8768	2157	4573	4262
Q2	1608	664	528	765	9123	2578	3607	4597
Q3	1606	657	498	821	9087	2519	3374	4835
Q4	1690	885	637	934	9302	3333	4184	5408
1971 Q1	1656	591	699	764	8856	2180	4478	4309
Q2	1776	739	588	857	9141	2690	3664	4684
Q3	1796	709	547	914	9025	2550	3375	4924
Q4	1877	951	713	1065	9290	3351	4322	5754
1972 Q1	1856	667	758	894	9028	2318	4553	4694
Q2	1863	793	661	962	9067	2699	3902	5047
Q3	1902	818	647	1041	8965	2751	3788	5436
Q4	1993	1092	808	1249	9188	3591	4648	6446
1973 Q1	2048	823	850	1046	8959	2668	4807	5392
Q2	2141	896	717	1088	9227	2818	4151	5525
Q3	2246	907	676	1176	9472	2765	3944	5862
Q4	2316	1234	886	1415	9462	3628	4801	6849
1974 Q1	2293	883	939	1204	8629	2483	4632	5572
Q2	2454	1043	894	1298	9140	2776	4017	5673
Q3	2577	1107	916	1435	9352	2857	3988	5984
Q4	2704	1465	1217	1761	9349	3617	4998	6918
1975 Q1	2772	1099	1310	1483	8749	2615	4789	5461
Q2	3043	1246	1168	1586	9018	2903	3965	5510
Q3	3221	1249	1128	1675	9316	2821	3723	5629
Q4	3277	1612	1513	2025	9397	3556	4767	6577
1976 Q1	3323	1174	1656	1697	8822	2532	5012	5313
Q2	3487	1369	1324	1749	9155	2876	3929	5345
Q3	3711	1401	1284	1904	9524	2884	3747	5679
Q4	3938	1853	1793	2362	9365	3674	4937	6781

1977 Q1	3883	1326	1879	1935	8660	2540	4996	5309
Q2	4129	1479	1611	2017	9081	2773	4143	5292
Q3	4159	1641	1496	2233	9086	2973	3942	5731
Q4	4425	2184	1909	2761	9720	3858	4858	6907
1978 Q1	4322	1552	2048	2286	9061	2714	5192	5560
Q2	4469	1773	1635	2459	9069	3040	4227	5829
Q3	4765	1962	1531	2619	9570	3313	3974	6090
Q4	4817	2545	2009	3246	9517	4207	4971	7419
1979 Q1	4793	1784	2390	2619	9015	2909	5691	5764
Q2	5104	2158	1864	2929	9411	3458	4304	6213
Q3	5482	2234	2002	3118	9833	3387	3995	6131
Q4	5609	2992	2590	3917	9787	4466	5002	7456
1980 Q1	5652	2051	2909	3246	9381	3047	5449	5861
Q2	5891	2295	2366	3434	9511	3290	4059	5935
Q3	6046	2396	2442	3546	9670	3373	4001	5975
Q4	6066	3131	3284	4375	9533	4354	5233	7183
1981 Q1	5894	2153	3480	3572	9204	2978	5333	5805
Q2	6095	2308	2998	3723	9329	3183	4254	5949
Q3	6417	2395	2934	3830	9672	3283	3921	6028
Q4	6540	3299	4010	4678	9644	4490	5300	7295
1982 Q1	6359	2263	4009	3846	9262	3051	5340	5879
Q2	6732	2479	3340	3990	9538	3299	4266	5991
Q3	6573	2603	3273	4194	9455	3430	4071	6215
Q4	6826	3580	4405	5182	9687	4667	5230	7550
1983 Q1	6477	2449	4518	4152	9145	3177	5360	6022
Q2	6887	2805	3783	4396	9506	3591	4435	6169
Q3	7210	2921	3427	4564	9841	3712	3971	6333
Q4	7487	3945	4492	5652	10090	4961	5166	7772
1964 Q1	6947	2600	4910	4490	9202	3279	5658	6087
Q2	7431	3000	3858	4881	9526	3727	4387	6511
Q3	7351	3199	3518	4988	9465	3935	4012	6606
Q4	7545	4369	4687	6278	9732	5320	5242	8183
1985 Q1	7157	2981	5281	4991	9045	3608	5844	6372
Q2	7636	3431	4276	5427	9593	4094	4574	6789
Q3	7788	3607	3967	5615	9786	4256	4277	6951
Q4	8076	4893	5054	7021	9978	5657	5496	8600
1986 Q1	7599	3364	5644	5652	9310	3930	6429	6873
Q2	8166	3841	4092	6105	9912	4468	4893	7358
Q3	8312	4035	3681	6310	10131	4657	4432	7552
Q4	8497	5406	4802	7854	10257	6114	5666	9252
1987 Q1	7920	3568	5654	6244	9390	4065	6543	7393
Q2	8505	4089	4056	6790	10049	4659	4759	7929
Q3	8808	4299	3810	7057	10492	4918	4500	8159
Q4	9169	5892	5108	8844	10690	6562	6069	10009

1988 Q1	8751	3834	5458	7068	10056	4312	6510	8017
Q2	9097	4386	4212	7614	10390	4808	4922	8560
Q3	9234	4570	4108	7892	10539	5027	4733	8753
Q4	9409	6233	5513	10062	10557	6633	6317	10978
1989 Q1	9130	4129	5420	7977	10053	4458	6139	8674
Q2	9907	4682	4814	8455	10793	4883	5173	9041
Q3	9806	4729	4283	8634	10621	4947	4635	9131
Q4	10300	6307	5943	11205	10780	6374	6388	11640
1990 Q1	9760	4377	5817	8869	9915	4556	6199	9095
Q2	10684	5004	5044	9336	10716	4994	5135	9404
Q3	10485	5088	4944	9397	10472	5127	4777	9363
Q4	10888	6407	6617	11964	10714	6199	6311	11704
1991 Q1	10318	4409	6820	9120	9882	4426	6638	8771
Q2	11223	5047	5793	9789	10665	4865	5355	9131
Q3	11076	5173	5188	9847	10571	5098	4665	9088
Q4	11427	6783	7154	12542	10751	6428	6493	11560
1992 Q1	10634	4391	7077	9659	9852	4335	6468	8772
Q2	11621	5135	5681	10408	10834	4945	5058	9346
Q3	11287	5352	5285	10358	10728	5293	4761	9264
Q4	11701	7219	7356	12736	10970	6882	6602	11357
1993 Q1	11040	4708	7220	10327	10159	4656	6479	9113
Q2	11708	5395	5858	10952	10832	5164	5071	9613
Q3	11445	5660	5498	11004	10614	5520	4774	9673
Q4	12041	7765	7560	13516	11196	7325	6697	11826
1994 Q1	11239	5032	7505	10719	10416	4902	6662	9371
Q2	11825	5725	6248	11376	10901	5462	5200	9878
Q3	11734	6010	5673	11423	10853	5853	4639	9929
Q4	12250	8072	7431	14138	11196	7633	6219	12237
1995 Q1	11748	5132	7869	11041	10545	5019	6536	9516
Q2	12424	5973	6114	11662	11091	5678	4927	9934
Q3	12212	6153	5585	11563	10856	5991	4477	9783
Q4	12466	8543	7624	14637	11027	8052	6269	12337

C. Values of the population parameters

The model is $y_t = Bx_t + u_t$. The population parameters for DGP1, DGP1*, DGP2, DGP2* and DGP3 are given below. To each population parameter a superscript is added which denotes the aggregation level.

1. Values of the population parameters of B for DGP1, DGP1*, DGP2, DGP2*

$$B^{(1)} = (\ -0.0003 \quad 0.1674 \quad -0.0214 \quad 0.0214\ )$$

$$B^{(2)} = \begin{pmatrix} -0.0001 & 0.1421 & -0.0162 & -0.0013 & 0.0175 \\ -0.0001 & 0.1833 & 0.0349 & -0.1083 & 0.0734 \end{pmatrix}$$

$$B^{(3)} = \begin{pmatrix} -0.0009 & 0.2016 & 0.0050 & 0.0088 & -0.0470 & 0.0331 \\ -0.0000 & 0.1728 & 0.0368 & -0.1127 & 0.0281 & 0.0478 \\ -0.0007 & 0.2780 & 0.0037 & 0.0387 & -0.0674 & 0.0250 \end{pmatrix}$$

$$B^{(4)} = \begin{pmatrix} -0.0009 & 0.1984 & 0.0037 & 0.0119 & -0.0431 & 0.0119 & 0.0156 \\ 0.0004 & 0.1430 & 0.0238 & -0.1019 & 0.0653 & 0.0271 & -0.0143 \\ -0.0002 & 0.2350 & -0.0143 & 0.0285 & -0.0624 & -0.0225 & 0.0707 \\ -0.0001 & 0.1477 & -0.0090 & 0.0533 & 0.0006 & -0.0115 & -0.0334 \end{pmatrix}$$

2. Values of the population parameters of R and Σ for DGP1, DGP1*

The VAR(1) error process is defined as $u_t = Ru_{t-1} + \epsilon_t$ with $\epsilon_t \sim i.i.d.N(O, \Sigma)$.

$$10^5 \cdot \Sigma^{(1)} = 0.1471$$

$$R^{(1)} = 0.2026$$

$$10^5 \cdot \Sigma^{(2)} = \begin{pmatrix} 0.1349 & -0.0230 \\ -0.0230 & 0.0954 \end{pmatrix}$$

$$R^{(2)} = \begin{pmatrix} 0.0832 & -0.3777 \\ -0.3281 & 0.0349 \end{pmatrix}$$

The eigenvalues of $R^{(2)}$ are 0.4119 and -0.2938.

$$10^5 \cdot \Sigma^{(3)} = \begin{pmatrix} 0.1215 & -0.0247 & -0.0165 \\ -0.0247 & 0.0965 & -0.0183 \\ -0.0165 & -0.0183 & 0.0848 \end{pmatrix}$$

$$R^{(3)} = \begin{pmatrix} 0.1180 & -0.1556 & 0.0405 \\ -0.3465 & 0.0246 & 0.0653 \\ 0.2689 & 0.1962 & 0.3177 \end{pmatrix}$$

The eigenvalues of $R^{(3)}$ are -0.2103, 0.3068 and 0.3638.

$$10^5 \cdot \Sigma^{(4)} = \begin{pmatrix} 0.1153 & -0.0339 & -0.0195 & -0.0430 \\ -0.0339 & 0.0815 & -0.0145 & -0.0001 \\ -0.0195 & -0.0145 & 0.0700 & -0.0179 \\ -0.0430 & -0.0001 & -0.0179 & 0.1068 \end{pmatrix}$$

$$R^{(4)} = \begin{pmatrix} 0.1547 & -0.1369 & 0.2050 & 0.0999 \\ -0.2638 & -0.0063 & 0.3899 & 0.3200 \\ 0.2435 & -0.0096 & 0.3353 & -0.0439 \\ 0.0670 & 0.4023 & -0.0663 & 0.1139 \end{pmatrix}$$

The eigenvalues of $R^{(4)}$ are $-0.4385, 0.4354 \pm 0.0636i$ and 0.1653.

3. Values of the population parameters of the vector MA(1) error process for DGP2, DGP2*

The vector MA(1) error process is defined as $u_t^* = \epsilon_t + \Theta\epsilon_{t-1}$ with $\epsilon_t \sim i.i.d.N(O, \Sigma^*)$.

$$10^5 \cdot \Sigma^{*(1)} = 0.74670$$

$$\Theta^{(1)} = 0.21163$$

$$10^5 \cdot \Sigma^{*(2)} = \begin{pmatrix} 0.60732 & -0.06521 \\ -0.06521 & 0.40923 \end{pmatrix}$$

$$\Theta^{(2)} = \begin{pmatrix} 0.12304 & -0.47195 \\ -0.38484 & 0.07939 \end{pmatrix}$$

$$10^5 \cdot \Sigma^{*(3)} = \begin{pmatrix} 0.61081 & -0.09394 & -0.10328 \\ -0.09394 & 0.42127 & -0.06162 \\ -0.10328 & -0.06162 & 0.36851 \end{pmatrix}$$

$$
\Theta^{(3)} = \begin{pmatrix} 0.13334 & -0.19544 & 0.06712 \\ -0.36157 & 0.05128 & 0.06247 \\ 0.29301 & 0.20358 & 0.38842 \end{pmatrix}
$$

$$
10^5 \cdot \Sigma^{*(4)} = \begin{pmatrix} 0.55427 & -0.19133 & -0.16021 & -0.15247 \\ -0.19133 & 0.22081 & -0.03254 & 0.02464 \\ -0.16021 & -0.03254 & 0.25783 & -0.03401 \\ -0.15247 & 0.02464 & -0.03401 & 0.42464 \end{pmatrix}
$$

$$
\Theta^{(4)} = \begin{pmatrix} 0.14204 & -0.30929 & 0.29881 & 0.09748 \\ -0.37821 & -0.25520 & 0.33495 & 0.38718 \\ 0.50772 & 0.21969 & 0.74218 & -0.03530 \\ 0.36106 & 1.06069 & 0.10146 & 0.20639 \end{pmatrix}
$$

4. Values of the population parameters of $R_x, \Sigma_\varsigma, \mu_x$ for DGP1*, DGP2*

$$R_x^{(1)} = \begin{pmatrix} 0.0300 & 0.1600 & -0.2769 \\ 1.1370 & 0.6753 & 0.3196 \\ 1.1631 & 0.2939 & 0.7578 \end{pmatrix}$$

The eigenvalues of $R_x^{(1)}$ are $0.1436, 0.8140, 0.5054$.

$$10^3 \cdot \Sigma_\varsigma^{(1)} = \begin{pmatrix} 0.2317 & -0.1013 & -0.2638 \\ -0.1013 & 0.4552 & 0.3223 \\ -0.2638 & 0.3223 & 0.6397 \end{pmatrix}$$

$$\mu_x^{(1)} = \begin{pmatrix} 0.0163 \\ 0.0555 \\ 0.0569 \end{pmatrix}$$

$$R_x^{(2)} = \begin{pmatrix} 0.1005 & 0.2001 & 0.0014 & -0.2978 \\ 0.9143 & 0.5711 & -0.1928 & 0.5566 \\ 0.9332 & -0.2452 & 0.2569 & 1.0091 \\ 0.9514 & 0.3835 & -0.1723 & 0.7672 \end{pmatrix}$$

The eigenvalues of $R_x^{(2)}$ are $0.7101, 0.2283, 0.6272, 0.1301$.

$$10^3 \cdot \Sigma_\varsigma^{(2)} = \begin{pmatrix} 0.2290 & -0.0921 & -0.3120 & -0.2279 \\ -0.0921 & 0.4108 & 0.2354 & 0.3053 \\ -0.3120 & 0.2354 & 0.8898 & 0.5073 \\ -0.2279 & 0.3053 & 0.5073 & 0.6378 \end{pmatrix}$$

$$\mu_x^{(2)} = \begin{pmatrix} 0.0167 \\ 0.0555 \\ 0.0633 \\ 0.0536 \end{pmatrix}$$

$$R_x^{(3)} = \begin{pmatrix} 0.0473 & 0.1649 & -0.0053 & 0.0160 & -0.2530 \\ 0.5915 & 0.2325 & -0.1825 & 0.8516 & 0.1598 \\ 0.8498 & -0.3378 & 0.2938 & 0.5028 & 0.6142 \\ 0.6179 & 0.3316 & -0.0540 & 0.7045 & -0.0800 \\ 0.7778 & 0.0469 & -0.2056 & 0.7378 & 0.5465 \end{pmatrix}$$

The eigenvalues of $R_x^{(3)}$ are $-0.0565, 0.8042, 0.1642, 0.4563 \pm 0.1806i$.

$$10^3 \cdot \Sigma_\varsigma^{(3)} = \begin{pmatrix} 0.2330 & -0.1120 & -0.3280 & -0.0710 & -0.3171 \\ -0.1120 & 0.3168 & 0.2175 & 0.1726 & 0.2693 \\ -0.3280 & 0.2175 & 0.9293 & 0.1687 & 0.6444 \\ -0.0710 & 0.1726 & 0.1687 & 0.2539 & 0.2101 \\ -0.3171 & 0.2693 & 0.6444 & 0.2101 & 0.9702 \end{pmatrix}$$

$$\mu_x^{(3)} = \begin{pmatrix} 0.0165 \\ 0.0555 \\ 0.0633 \\ 0.0391 \\ 0.0604 \end{pmatrix}$$

$$R_x^{(4)} = \begin{pmatrix} 0.0185 & 0.1832 & -0.0252 & -0.0158 & -0.1295 & -0.1035 \\ 0.6411 & 0.1957 & -0.1272 & 0.9586 & 0.1145 & -0.0449 \\ 0.8632 & -0.3457 & 0.3154 & 0.5285 & 0.2622 & 0.3271 \\ 0.5816 & 0.3414 & -0.0526 & 0.7079 & -0.0556 & -0.0406 \\ 0.9881 & -0.0538 & -0.3311 & 0.8784 & 0.6429 & 0.0108 \\ 0.7370 & 0.0545 & -0.0279 & 0.8021 & 0.0262 & 0.3004 \end{pmatrix}$$

The eigenvalues of $R_x^{(4)}$ are $0.8232, -0.0986, 0.5042 \pm 0.2013i, 0.1620, 0.2859$

$$10^3 \cdot \Sigma_\varsigma^{(4)} = \begin{pmatrix} 0.2386 & -0.1130 & -0.3411 & -0.0797 & -0.4581 & -0.2298 \\ -0.1130 & 0.3164 & 0.2258 & 0.1808 & 0.3026 & 0.2481 \\ -0.3411 & 0.2258 & 0.9629 & 0.1808 & 0.8513 & 0.5433 \\ -0.0797 & 0.1808 & 0.1808 & 0.2589 & 0.2505 & 0.2122 \\ -0.4581 & 0.3026 & 0.8513 & 0.2505 & 2.1417 & 0.7392 \\ -0.2298 & 0.2481 & 0.5433 & 0.2122 & 0.7392 & 0.7355 \end{pmatrix}$$

$$\mu_x^{(4)} = \begin{pmatrix} 0.0167 \\ 0.0555 \\ 0.0633 \\ 0.0391 \\ 0.0702 \\ 0.0541 \end{pmatrix}$$

5. Values of the population parameters of $B, \Sigma^{**}, R_1, R_2, R_3, R_4$ for DGP3

$$B^{(1)} = \begin{pmatrix} -0.0027 & 0.2768 & -0.1098 & 0.1098 \end{pmatrix}$$

$$B^{(2)} = \begin{pmatrix} -0.0039 & 0.3136 & -0.0870 & -0.0004 & 0.0874 \\ -0.0002 & 0.2514 & 0.0323 & -0.0591 & 0.0268 \end{pmatrix}$$

$$B^{(3)} = \begin{pmatrix} -0.0044 & 0.3352 & -0.0836 & -0.0144 & 0.0263 & 0.0716 \\ 0.0001 & 0.2430 & 0.0270 & -0.0571 & -0.0031 & 0.0333 \\ -0.0002 & 0.2003 & 0.0137 & 0.0043 & -0.0210 & 0.0031 \\ 0.0045 & 0.2214 & 0.0430 & 0.0672 & -0.0022 & -0.1080 \end{pmatrix}$$

$$10^5 \cdot \Sigma^{**(1)} = 2.6917$$

$$10^5 \cdot \Sigma^{**(2)} = \begin{pmatrix} 2.5532 & -0.6350 \\ -0.6350 & 1.4149 \end{pmatrix}$$

$$10^5 \cdot \Sigma^{**(3)} = \begin{pmatrix} 2.3821 & -0.5730 & -1.2045 \\ -0.5730 & 1.2684 & -0.8807 \\ -1.2045 & -0.8807 & 3.1139 \end{pmatrix}$$

$$R_1^{(1)} = 0.3341 \quad R_2^{(2)} = -0.0011 \quad R_3^{(3)} = 0.0466 \quad R_4^{(4)} = -0.4141$$

$$R_1^{(2)} = \begin{pmatrix} 0.4224 & 0.2475 \\ -0.1146 & 0.2164 \end{pmatrix} \qquad R_2^{(2)} = \begin{pmatrix} -0.0356 & -0.1592 \\ 0.0131 & 0.2788 \end{pmatrix}$$

$$R_3^{(2)} = \begin{pmatrix} 0.0892 & -0.0028 \\ 0.0557 & 0.0763 \end{pmatrix} \qquad R_4^{(2)} = \begin{pmatrix} -0.4567 & -0.0673 \\ 0.1774 & -0.0075 \end{pmatrix}$$

$$R_1^{(3)} = \begin{pmatrix} 0.4723 & 0.2144 & 0.0325 \\ -0.0553 & 0.3088 & 0.1499 \\ 0.0747 & -0.0115 & 0.2646 \end{pmatrix} \qquad R_2^{(3)} = \begin{pmatrix} 0.1830 & 0.1758 & 0.2885 \\ -0.1565 & 0.0531 & -0.1563 \\ 0.0634 & -0.0123 & 0.0483 \end{pmatrix}$$

$$R_3^{(3)} = \begin{pmatrix} -0.1327 & -0.3206 & -0.2657 \\ 0.2482 & 0.3396 & 0.2027 \\ -0.0782 & -0.0077 & 0.1380 \end{pmatrix} \quad R_4^{(3)} = \begin{pmatrix} -0.3043 & 0.0624 & 0.1819 \\ 0.1367 & -0.0601 & -0.0746 \\ 0.1224 & 0.0152 & -0.0808 \end{pmatrix}$$

D. Algorithm for the MA(1) parameters

An approximate algorithm is proposed for defining the population parameters of the vector MA(1) process for given numerical values of its variance-covariance matrix and the first autocovariance matrix. The MA(1) error process is defined by

$$u_t = \epsilon_t + \Theta\epsilon_{t-1} \quad \text{with } \epsilon_t \sim i.i.d.N(O, \Sigma^*) \quad .$$

The variance-covariance matrix of u_t is given by (see chapter 4.4. example 4.2)

$$\Sigma_{u,0}^* := E(u_t u_t') = \Sigma^* + \Theta\Sigma^*\Theta' \quad .$$

The autocovariance matrix of order one is given by

$$\Sigma_{u,1}^* := E(u_t u_{t-1}') = \Theta\Sigma^* \quad .$$

Here we want to define a vector MA(1) error process which has approximately the same variance-covariance and first autocovariance matrix as the VAR(1) error process for DGP1/DGP1*. The variance-covariance matrix of the VAR(1) process was denoted by $\Sigma_{u,0}$ and the first autocovariance matrix by $\Sigma_{u,1}$.

For given numerical values of $\Sigma_{u,0}$ and $\Sigma_{u,1}$, the population parameters of the vector MA(1) process Σ^* and Θ are chosen such that

$$\Sigma_{u,0} \approx \Sigma_{u,0}^* = \Sigma^* + \Theta\Sigma^*\Theta' \tag{D.1}$$

and

$$\Sigma_{u,1} \approx \Sigma_{u,1}^* = \Theta\Sigma^* \quad . \tag{D.2}$$

The two equations above cannot be solved analytically for Σ^* and Θ simultaneously. Here the values of Σ^* and Θ are approximated as follows: for given Θ, one can solve for Σ^* in equation (D.1) by using the vec operator:

$$\text{vec } \Sigma_{u,0} = \text{vec } \Sigma^* + (\Theta \otimes \Theta)\text{vec } \Sigma^*$$

and by this

$$\text{vec } \Sigma^* = \left(I_n^2 + (\Theta \otimes \Theta)\right)^{-1} \text{vec } \Sigma_{u,0} \quad . \tag{D.3}$$

For given Σ^*, it follows from equation (D.2) that

$$\Theta = \Sigma_{u,1} \Sigma^{*-1} \quad . \tag{D.4}$$

For finding the numerical values of Σ^* and Θ, we propose to iterate on (D.3) and (D.4). For the first step of the algorithm, the starting values of Θ are set to the zero matrix in order to calculate (D.3). The last step of the algorithm consists in calculating equation (D.4). By this, $\Sigma_{u,1} = \Theta \Sigma^*$ holds exactly (apart from numerical inaccuracies). In order to check if the algorithm converges, the iteration process is ended if all elements of the matrix $(\Sigma_{u,0} - \Sigma^* + \Theta \Sigma^* \Theta')$ are smaller than 10^{-10}. Note that $\Sigma_{u,0}$ is of order 10^{-5}. It should be emphasised that this algorithm is heuristic and does not necessarily converge. In the simulation study, the number of equations n is varied between 1 and 4 for DGP2 and DGP2*. For $n = 1, 2, 3$, the algorithm works very well and convergence is achieved after a few steps. For $n = 4$, problems occurred. The quality of approximation does not improve with the number of iterations. The reason for this seems to be the following. Recall that the first autocovariance matrix of the VAR(1) process u_t was given by $\Sigma_{u,1} = R\Sigma_{u,0}$. For $n = 4$, the matrix R has complex eigenvalues while for $n = 1, 2, 3$ all the eigenvalues of the R matrices are real. Therefore, the following brute force procedure is used. For $n = 4$, we iterated 10000 times on the equations (D.3) and (D.4) and simply chose the "best" solution. The approximate solution is close to the exact solution to the order of 10^{-7}.

List of Figures

List of Tables

References

Amemiya, T. (1985) Advanced Econometrics. Harvard University Press, Cambridge Massachusetts

it Anderson, T.W. (1971) The Statistical Analysis of Time Series. Wiley, New York

Anderson, T.W. (1984) An Introduction to Multivariate Analysis, 2nd edn. Wiley, New York,

it Andrews, D. (1991) Heteroskedasticity and Autocorrelation consistent Covariance Matrix Estimation. Econometrica 59:817-858

Andrews, D. and Monahan, C. (1992) An improved Heteroskedasticity and Autocorrelation consistent Covariance Matrix Estimator. Econometrica 60:953-966

Barnard, G.A. (1963) Contribution to Discussion. Journal of the Royal Statistical Society B 25:294

Barten, A.P. (1969) Maximum Likelihood Estimation of a Complete System of Demand Equations. European Economic Review 1:7-73

Barten, A.P. and Böhm, V. (1982) Consumer Theory. In: Arrow K.J. and Intriligator, M.D. (Eds.) Handbook of Mathematical Economics. North-Holland, Amsterdam, vol. 2, ch. 9

Bera, A. (1982) A Note on testing Demand Homogeneity. Journal of Econometrics 18:291-294

Berndt, E.R. and Savin, N.E. (1975) Estimation and Hypothesis Testing in Singular Equation Systems with autoregressive Disturbances. Econometrica 43:937-957

Bewley, R. (1986) Allocation Models. Series on Econometrics and Management Sciences. Ballinger, Cambridge Massachusetts, vol. 6

Central Statistical Office (CSO) Economic Trends. Annual Supplement 1996

Central Statistical Office (CSO) Statistical Yearbooks. 1977, 1991, 1996

Central Statistical Office (CSO) (1985) United Kingdom National Accounts: Sources and Methods

Cribari-Neto, F. and Zarkos, S. (1997) Finite-Sample Adjustments for Homogeneity and Symmetry Tests in Systems of Demand Equations: A Monte Carlo Evaluation. Computational Economics 10:337-351

Davidson, R. and MacKinnon, J.G. (1992) Regression based Methods for using Control Variates in Monte Carlo Experiments. Journal of Econometrics

54:203-222

Davidson, R. and MacKinnon, J.G. (1993) Estimation and Inference in Econometrics. Oxford University Press, Oxford, ch. 21,

Davidson, R. and MacKinnon, J.G. (1998) Graphical Methods for investigating the Size and Power of Hypothesis Tests. The Manchester School 66:1-26

Davies, R.B. (1980) The Distribution of a Linear Combination of random χ^2 Variables: Algorithm AS 155. Applied Statistics 29:323-333

Deaton, A.S and Muellbauer J. (1980b) Economics and Consumer Behaviour. Cambridge University Press, Cambridge

Den Haan, W.J. and Levin. A. (1997) A Practioner's Guide to Robust Covariance Matrix Estimation. In: Maddala, G.S. and Rao, C.R. (Eds.) Handbook of Statistics. North-Holland, Amsterdam, vol. 15, ch. 12

Deschamps, P.J. (1993) Joint Tests for Regularity and Autocorrelation in Allocation Systems. Journal of Applied Econometrics 8:195-211

Deschamps, P.J. (1988) A Note on Maximum Likelihood Estimation of Allocation Systems. Computational Statistics and Data Analysis 6:109-112

Dwass, M. (1957) Modified Randomization Tests for nonparametric Hypotheses . Annals of Mathematics and Statistics 28:181-187

Giles, D. and Scott, M. (1992) Some Consequences of using the Chow Test in the Context of autocorrelated Disturbances. Economics Letters 38:145-150

GAUSS, The Gauss System Version 3.2.12: User Manuals: Aptech Systems, Washington

Hall, P. and Titterington, D.M. (1989) The Effect of Simulation Order on Level Accuracy and Power of Monte Carlo Tests. Journal of the Royal Statistical Society, Series B. 51:459-467

Hamilton, J.D. (1994) Time Series Analysis. Princeton University Press, Princeton New Jersey

Hansen, B. (1992) Consistent Covariance Estimation for dependent heterogeneous Processes. Econometrica 60:967-972

Hendry, D.F. (1984) Monte Carlo Experimentation in Econometrics. In: Griliches, Z. and Intriligator, M.D. (Eds.) Handbook of Econometrics. North-Holland, Amsterdam, vol. 2, ch. 16

Hogg, R.V. and Craig, A.T. (1978) Introduction to Mathematical Statistics, 5th edn. Prentice Hall

Horowitz, J.L. (1996) Bootstrap Methods in Econometrics: Theory and numerical Performance. In: Kreps, D.M. and Wallis, K.F. (Eds.) Advances in Economics and Econometrics: Theory and Applications Seventh World Congress. Cambridge University Press, Cambridge

Imhof, J. (1961) Computing the Distribution of Quadratic Forms in normal Variables. Biometrika 48:419-426

Laitinen, K. (1978) Why is Demand Homogeneity so often rejected?, Economics Letters 1:187-191

Magnus, J.R. and Neudecker, H. (1988) Matrix differential Calculus with Applications in Statistics and Econometrics. John Wiley & Sons, Chichester

Mizon, G.E. and Hendry, D.F. (1980) An Empirical Application and Monte Carlo Analysis of Tests of Dynamic Specification. Review of Economic Studies 47:21-45.

Pesaran, M.H. and Pesaran, B. (1991) MICROFIT 3.0 An Interactive Econometric Software Package. User Manual

Pesaran, M.H. and Pesaran, B. (1997) MICROFIT 4.0 An Interactive Econometric Software Package. User Manual

Ray, S., Ravikumar, B. and Savin, E. (1998) Robust Wald Tests in SUR Systems with Adding up Restrictions: An Algebraic Approach to Proofs of Invariance. Working Paper. University of Iowa

Rilstone, P. (1993) Some Improvements for bootstrapping Regression Estimators under first-order serial Correlation. Economics Letters 42:335-339

Veall, M.R. (1998) Applications of the Bootstrap in Econometrics and Economic Statistics. In: Giles, D. and Ullah, A. (Eds.) Handbook of Applied Statistics. M. Decker, New York, ch. 12

White, H. (1984) Asymptotic Theory for Econometricians. Academic Press, Orlando

Index

Lecture Notes in Economics and Mathematical Systems

For information about Vols. 1–295
please contact your bookseller or Springer-Verlag